Foreword

This publication, *Petrography in Cementitious Materials*, contains papers presented at the symposium of the same name, held in Atlanta, GA on 23 June 1993. The symposium was sponsored by ASTM Committee C-9 on Concrete and Concrete Aggregates and ASTM Committee C-1 on Cement. Sharon M. DeHayes of the California Portland Cement Company in Glendora, CA and David Stark of Construction Technology Laboratories in Skokie, IL presided as symposium chairpeople and are editors of the resulting publication.

Contents

Overview

On 23 June, 1993, ASTM Committees C-1 on Cement and C-9 on Concrete and Concrete Aggregates, cosponsored a symposium on the Petrography of Cementitious Materials. Papers in this symposium focused primarily on cementitious materials in portland cement concrete as characterized by optical and electron microscopy and by other analytical techniques. The intent of this symposium was to complement a previous symposium (STP 1061) held on 26 June, 1989 at the ASTM meeting in St. Louis, MO entitled Petrography Applied to Concrete and Concrete Aggregates. It was hoped the two symposia together would demonstrate the wide ranging potential of petrography as a tool in forensic, as well as in fundamental investigations of cement and concrete properties and performance.

An underlying theme in the present symposia papers is the description of techniques used to characterize the properties of cementitious materials and hardened concrete. The paper by Ahmed describes techniques and the need for obtaining high quality surfaces for optical microscopical observation of cement clinker and concrete. Without quality surfaces, observations and conclusions are limited and can be misleading. Campbell describes the use of a transmitted light technique, commonly referred to as the Ono Method, for characterizing phase composition, specifically belite, as it relates to quality control in the production of portland-cement clinker. The advantage of combining optical microscopy with scanning electron microscopy and energy dispersive X-ray analysis to characterize clinker production is demonstrated in the paper by Sarkar and Samet. The paper by Weigand reports results of inter- and intra-laboratory studies using optical microscopy to determine phase composition of portland-cement clinker. This study was sponsored by ASTM Committee C01.23 on Compositional Analysis.

Bentz and Stutzman, and Stutzman, report on the use of computer modeling and scanning electron microscopy to determine and describe the phase composition and microstructure of clinker and portland cement. The use of these techniques in developing a two-dimensional model of cement-hydration processes is described in one of the papers.

Three papers in the symposium focus on characteristics of hydrated cement paste as they may affect the performance of concrete. The Wirgot and Van Cauwelaert paper deals with determining the water-cement ratio of concrete using a fluorescent epoxy resin as the adhesive for petrographic thin sections. Image analysis and statistical analyses are used to obtain final results. The paper by Cahill et al. describes the use of image analysis to determine the entrained air content of hardened concrete. The Lankard et al. paper describes the alteration of concrete located adjacent to electrode/concrete interfaces in structures that have been under cathodic protection.

All of the papers in this symposium are timely in their technical content and should familiarize the reader with recent developments, primarily in the application of microscopic analytical techniques to concrete performance. It also is hoped that these symposia papers, together with those in the previous symposium held in 1989, will lead to the wider use of petrographic techniques in maintaining quality control of concrete products and to clearer resolution of durability problems encountered in the field.

Sharon M. DeHayes
California Portland Cement Company, Glendora, CA; Symposium chairperson and editor.

David Stark
Construction Technology Laboratories, Skokie, IL; Symposium chairperson and editor.

Wase U. Ahmed*

PETROGRAPHIC METHODS FOR ANALYSIS OF CEMENT CLINKER AND CONCRETE MICROSTRUCTURE

REFERENCE: Ahmed, W. U., **"Petrographic Methods for Analysis of Cement Clinker and Concrete Microstructure,"** Petrography of Cementitious Materials, ASTM STP 1215, Sharon M. DeHayes and David Stark, Eds., American Society for Testing and Materials, Philadelphia, 1994.

ABSTRACT: Petrography has long been used to reveal the microstructure of ores and minerals. Its use in the cement industry however is fairly new. Petrographic investigation can reveal various phases of cement. These provide an understanding of the burning process in the kiln, along with equally important factors which govern the quality of the cement, such as mortar strength, weather resistance and grindability. To obtain this information, the clinker or the concrete specimen must be prepared (polished) in a specific manner so that it can be analyzed under an optical microscope. This paper describes the various methods of specimen preparation which are used for microscopic investigation and application of improved preparation equipment is emphasized.

KEYWORDS: Clinker, concrete, microscopy, image analysis, petrography, microstructure, specimen preparation

INTRODUCTION

Concrete, which has been used as a construction material for more than a century, is still the most popular building material, and no other material is used in as large a quantity as concrete. The low cost, great compressive strength, and durability makes concrete an ideal construction material for highways, bridges, buildings, homes, and countless other structures.

Cement, which is an important component of concrete, is the fine powder which when reacted with water bonds together other components such as sand and aggregates. It is produced by mixing limestone or other lime-bearing materials with silica-bearing materials such as clays. The mixture is heated to a temperature of approximately 1500°C in a kiln. When the material comes out of the kiln, it is in the form of nodules

* Mineralogist
BUEHLER LTD., 41 Waukegan Road, Lake Bluff, IL USA 60044

which may vary in size and are called "clinkers". The clinker is ground with gypsum to a fine powder and becomes cement.

Various aspects of concrete performance such as strength, weather, and chemical resistance, reactivity, etc., depend on the behavior of cement. Therefore, the quality of cement used in concrete is of utmost importance, and the microscopic examination is a valuable tool that provides important information to help control the manufacturing process.

Purpose of Microscopical Analysis

The properties of any material are a direct consequence of its chemistry and microstructure. Analysis by microscopy reveals the microstructure thereby providing an in-depth understanding of the cement properties.

The microscope when used properly is an indispensable laboratory tool used to obtain information on the mineralogy of cement clinker, homogeneity of the mix, shape and size of clinker phases, and alteration products. Other equally important properties such as porosity, which is almost always present in a cement clinker, determines how easy or difficult it is to grind the clinker. Since the grinding process consumes a significant amount of energy, analysis of clinker porosity by optical microscopy or electronic image analysis means can be very helpful in optimizing the porosity in manufacturing thereby reducing the production cost. As reported by Gouda, "Clinker grinding consumes about one-third of the power which is required to produce cement."[1]

Clinker consists of four principle phases: tricalcium silicate called alite (C_3S), dicalcium silicate called belite (C_2S), tricalcium aluminate called aluminate (C_3A), and tetra calcium aluminoferrite (C_4AF) called ferrite. The shape and size of alite and belite indicates the kiln conditions existing during the manufacturing process. Consequently by analyzing the data obtained by microscopy, production variables can be adjusted to make a product of good and consistent quality. However, for microscopic examination, a clinker specimen must be properly prepared to minimize obtaining erroneous results in the analysis.

METHODS OF SAMPLE PREPARATION FOR ANALYSIS

There are two types of specimens which are analyzed microscopically. These include polished sections and thin sections:

Polished Sections: This is the most popular and prevalent method for analysis in the cement industry. These specimens are easy to prepare and do not require knowledge of

[1]Gouda, George R., "Effective Clinker Composition on Grindability", Cement and Concrete Research, vol. 9, 1979.

mineralogy. Features such as shape and size of alite and belite, phase homogeneity, matrix composition, porosity and alteration products can be determined. The specimens are encapsulated in a thick plastic slug, polished, and examined under a reflected light microscope.

Thin Sections: Thin sections permit a more comprehensive analysis of the specimen, but they are more difficult to prepare. Thin sections are observed under the transmitted polarized light microscope, and analysis requires some knowledge of optical mineralogy.

In addition to revealing the features described under polished sections, additional features such as mineral contents of the samples can also be determined.

This paper outlines specimen preparation methods for polished and thin sections, and application of improved preparation equipment is given principle emphasis.

Preparation Sequence:

To prepare a polished or a thin section, the specimen must go through a series of preparation steps which generally consist of the following:

1. Sectioning
2. Impregnation and encapsulation
3. Planar grinding
4. Polishing

Figure 1. A precision diamond saw

Since specimen preparation techniques are intended to yield the true microstructure of a material, care must be exercised in all phases of sample preparation in order to avoid altering or hiding the true microstructure.

Sectioning: Sectioning generally produces significant deformation which can be minimized by using precision diamond saws and specially designed blades. The key concern in sectioning is to produce a surface which has a minimum of deformation and sectioning is performed within a short time.

The precision diamond saw, Figure 1, is ideal for sectioning clinkers and concrete which are prone to grain plucking. The high rotational speed and higher load keeps the cutting time short. Because of the wide range of blades which are available for this saw, it is easy to select a suitable blade for a specific application. Figure 1, A & B shows the difference in deformation produced in a brittle materials when sectioned with two different types of blades.

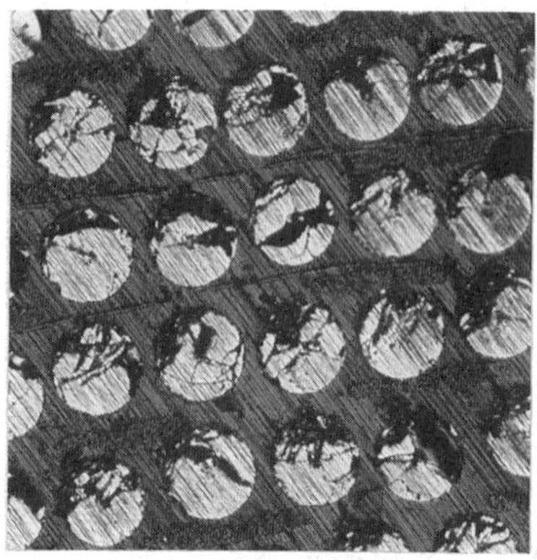

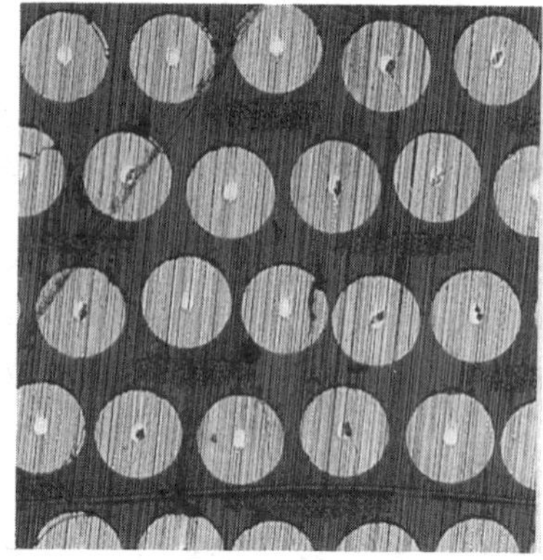

Figure 1A. A composite cut with a blade designed for brittle materials

Figure 1B. A composite cut with a general-purpose blade

By selecting a proper grit size of diamond abrasive and the composition and concentration of the bond, a series of blades have been developed to match specific sectioning requirements. For example, if determination of porosity in a sample is the prime concern, the sample must be sectioned in such a way that the sectioning itself does not induce porosity due to the grain pullout. In this application, a fine diamond blade with the correct bond will ensure the least damage to the sample.

A saw with adjustable speed allows the operator to select the optimum speed for a given material. Generally a higher speed and force results in a faster cutting time, Figure 2.

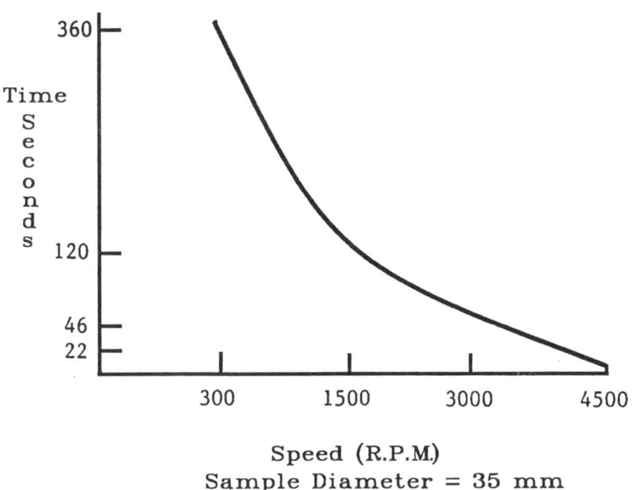

Figure 2.

To eliminate hydration which can be caused by the use of water-base coolants, other coolants such as straight propylene glycol can be used.

Impregnation and Encapsulation

Most clinkers and concrete specimens exhibit porosity and should be impregnated with a suitable medium such as epoxy to fill the pores and cracks. If the pores and cracks are not filled, they enlarge in grinding and become sites for entrapment of abrasives, lubricants, and solvents used for cleaning and etching. Failure to impregnate the sample can also cause cross contamination of abrasive and staining of the specimen. Although many casting resins can be used for impregnation, epoxies are ideal for this purpose because they bond well, display low shrinkage, and are not affected by most solvents and etchants which are routinely used in sample preparation. Epoxies or other plastics used for impregnation should have low viscosity to facilitate better penetration. Epoxies which display viscosity ranging from 100 - 300 centi-poise are ideal for impregnation.

If microscopic analysis has to be performed quickly, an acrylic encapsulating material which cures in about 6 minutes can be used. However, because of its higher viscosity and rapid setting time, it should not be used for impregnation purposes.

Planar Grinding, Sample Integrity, and Polishing (grinding and polishing)

A. Grinding:

Grinding and polishing can be accomplished either by hand on a rotating wheel or by a semiautomatic polishing device. The advantages of using a semiautomatic polishing device are numerous. When the preparation parameters such as time, force, and wheel speed are controlled, even a novice operator can repeatedly produce well-polished specimens within a short time. Most semiautomatic polishing systems are capable of preparing multiple specimens simultaneously and rapidly making it possible to extract specimens from the various zones of a kiln for analysis and then adjust the kiln conditions accordingly.

Although clinkers can be ground with loose abrasives such as silicon carbide or aluminum oxide, loose abrasives have a tendency to seep into the pores and are difficult to remove. Fixed abrasives such as abrasive papers or diamond grinding discs are easier to use, and specimens can be cleaned easily minimizing cross contamination. To eliminate hydration (if it is of concern), a 50:50 mixture of propylene glycol and alcohol or an oil-base lubricant can be used in grinding.

A typical grinding sequence can be as follows:

REFLECTED LIGHT SAMPLE PREPARATION SEQUENCE
For Cement Clinkers and Concrete

PRELIMINARIES		
Section	Clean	Vacuum Impregnate and/or Encapsulate
PLANAR GRIND		
Abrasive/Grit Size	**Wheel Covering**	**Lubricant**
SiC 400 and 600 grit (clinker)	Abrasive Paper	50% Propylene Glycol -- 50% Alcohol or Lapping Oil
SiC 240 - 1000 grit (Concrete) 70 - 30 micron	Cast Iron Lap Diamond disc	50% Propylene Glycol -- 50% Alcohol or Lapping Oil

B. Sample Integrity (polishing):

The purpose of polishing a specimen is to remove the deformation induced in the grinding process and is accomplished by abrading the surface with fine abrasives progressively decreasing to sub-micron size. Because this is the last step in the preparation sequence, any deformation remaining in the specimen will appear when the specimen is observed under the microscope. It is therefore necessary to polish the specimen until all deformation is eliminated. The polishing time will vary depending on the size of abrasive used, force applied, time and speed of the wheel.

A general guideline which can be followed with ease is to observe the specimen surface under a microscope at the end of each step. The specimen surface is examined for pits and scratches. At any given stage, there may be present some pits and scratches which are always proportional to the size of the abrasive used in preparing the surface. When polishing ceases to improve the surface regardless of the time spent, the specimen should be cleaned thoroughly and moved to the next stage of polishing. This simple method to determine when to stop polishing will be very useful in deciding the time required for each step.

Unnecessary long time spent in polishing is not only wasteful but also results in producing relief in the specimen surface which is undesirable. Relief is the topographic difference between a soft and a hard mineral or a phase which makes microscopic observation difficult. This is because the reduced depth of field that occurs in optical microscopes, especially at higher magnification, makes focusing of the entire surface at the same time difficult.

A typical polishing sequence is as follows:

Stage	Integrity Surface (polishing surface)	Abrasive Size	Lubricant
Rough Polish	Hard Cloth or silk	6 micron diamond	A or B
Wash specimen thoroughly.			
Fine Polish	Soft Cloth	0.05 micron Alumina	A
Wash specimen thoroughly; rinse with alcohol and dry.			

A = 50:50 Ethylene glycol and alcohol
B = oil-base lubricant

At the end of each preparation stage, the specimen must be thoroughly cleaned with alcohol to prevent hydration. If hydration is not of concern, a water and liquid soap rinse can be used for cleaning. Water also acts as an etchant and brings out alite and belite. The following etchants are commonly used for clinker to distinguish various phases.

Etchants for Clinkers:

ETCHING REAGENTS AND STAINS FOR CEMENT CLINKERS

Structure to be Viewed	Etchant Composition	Comments
Aluminates and Free Lime	Distilled Water	Immerse the polished surface for 3 - 5 seconds.
Silicates	1% Solution of NH_4Cl in water OR: 1% HNO_3 in Ethyl Alcohol	Immerse the surface for 5 - 10 seconds.
Interstitial Glass	10% KOH in Water	For best results, the temperature of the etchant should be 30°C. Immerse for 10 - 15 seconds.
Stain to resolve Alite and Belite	1 gram NH_4NO_3 150 ml Isopropyl alcohol 20 ml Water 20 ml Ethyl Alcohol 10 ml Acetone Composition of Salicylic Acid Etchant: 0.2 grams Salicylic Acid 25 ml Isopropyl Alcohol 25 ml Water	Useful as an alite stain, 25 - 30 seconds. If followed by salycylic acid (30 seconds), will resolve alite and belite. Depending on the time, the color ranges from light brown to brown to purplish-brown to blue to blue-green to green to yellow-green. When alite is yellow-green, belite will probably be brown.

Thin Section Preparation

A thin section is indeed a very thin section that may vary in thickness from 30 microns down to 10 micron or less. Preparation of a thin section by hand requires skill, experience, and time that is generally not available to most laboratories. Because uniformity of the specimen thickness is of utmost concern, thin section preparation by

hand can be difficult. Hand grinding tend to favor one or the other side of a thin section which eventually results in uneven thickness of the specimen.
Automatic preparation systems reduce the skill, experience, and time required to prepare thin sections. A thin sectioning machine such as that shown in Figure 3 is capable of preparing thin sections accurately and rapidly. A precision micrometer is used to control the thickness of the thin section.

The specimen is ground in several passes. Initially 20 - 30 microns of material is removed in each pass but as the thickness of the specimen decreases close to 50 - 55 microns, only 5 - 10 microns of material is removed to prevent shattering of the grains.

When the thin section is approximately 40 - 45 micron thick, it is removed from the thin sectioning machine and is ground on a glass face wheel with 9.5 micron aluminum oxide until the desired thickness is obtained.

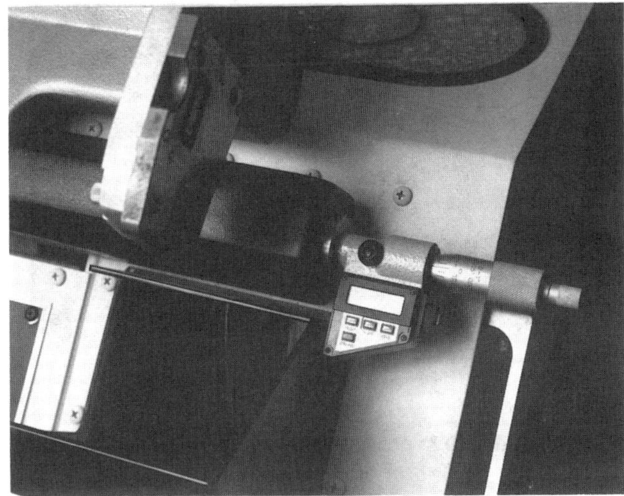

Figure 3. A thin sectioning machine for cutting and grinding thin sections

Once a thin section has been ground, it can be observed under the microscope, or it can be polished. A polished thin section can be examined either under a transmitted light microscope or a reflected light microscope.

To polish a specimen, it can be either prepared by hand or by an automatic polishing device such as shown in Figure 4 which is capable of preparing multiple specimens simultaneously. The slides are placed in a holder which is then placed under a cylinder which applies pressure on the holder. The rotation of the wheel causes the holders to rotate as well as spin on their own axis. This movement of the holders assures elimination of directional polishing effects which can occur in hand polishing.

To prevent relief in concrete specimens (between the paste and the aggregates), a hard napless cloth and use of a diamond abrasive is most effective.

Figure 4. A device to polish several thin sections simultaneously

Preparation of Ultra-thin Sections

Fine grain materials such as cement consist of crystals which can be finer than 30 microns in size. For thin section analysis, the conventional 30 micron thin section may be unsuitable. This is because a thin section which is 30 microns in thickness may contain several layers of the fine crystals which will make microscopic observation difficult and confusing. In order to observe such a specimen under the microscope, it is sometimes necessary to prepare ultra-thin sections which are much thinner than 30 microns. The conventional method of specimen preparation by hand is difficult because at this thickness even slight pressure applied in preparation can destroy the specimen. However, by using vibratory polishing it is possible to obtain ultra-thin sections more consistently. The vibratory polishing method is very gentle and removes material very slowly which is essential for preparing ultra-thin sections.

Most specimens can be left in the vibratory polisher for several hours without destroying the specimen. Generally specimens are polished with fine abrasives such as 1 micron diamond on a low-napped cloth. Coarser abrasives can also be used; however, specimens must be checked more frequently.

To polish the specimens, they are attached to a holder and placed inside a bowl containing the abrasive. Vibratory frequency is adjusted by a potentiometer until the holders start to move smoothly in the bowl. Because specimens are attached to individual holders, any specimen can be removed from the polishing bowl whenever required.

Method for Pore Measurement in Concrete Specimens especially for Image Analysis

Air entraining of concrete can improve its freeze/thaw resistance which is important in cold climate. The air voids must be present in the form of small pores well distributed throughout the hardened concrete. To analyze the air voids in a hardened concrete, the surface should be ground and polished.

The specimen is cut to a suitable size with a diamond saw and lapped using abrasive size ranging from 240 to 1000 grits Because the size of the air pores is important for analysis, petrographic preparation must not damage the pore structure by breaking the pore edges thus enlarging the pore size. To overcome this problem, it is necessary that the pores be impregnated with a suitable encapsulating material which may be accomplished by the use of a low viscosity epoxy. Because of the lower viscosity of the epoxy, it readily penetrates pores and cracks.

To enhance optical contrast between the pores and the matrix, which is essential for measurement, a white pigment (Titanium Dioxide) is added to the epoxy. Addition of the white pigment assists in distinguishing the pores very readily and is very useful for image analysis which is becoming a popular method for air-pore measurement. Also because the pores are filled with epoxy, the pore walls do not collapse in preparation which may happen if the pores are hollow causing enlargement of the pore size.

Method of Concrete Specimen Preparation to Increase Contrast for Image Analysis

- To prepare the specimen, vacuum impregnate it in epoxy to which Titanium Dioxide is added. To facilitate complete mixing of the pigment into the epoxy, heat the proper amount of resin to approximately 150° F, and then add the pigment. Mix it thoroughly to break any clumps. Dispense the pigmented resin according to the instructions supplied by the manufacturer.

- Depending on the surface finish of the concrete specimen, lap the surface down to 1000 grit silicon carbide on a cast iron wheel.
- Wash the sample thoroughly and dry it in an oven at 100 - 200°F for several hours.
- Wet a stamp pad with black India ink (do not over saturate) and press against the lapped surface several times to blacken the surface.
- Dry the specimen in the oven for several hours until the ink is completely dry.
- Re-grind the surface for as short a time as possible with a fine abrasive paper or 1000 grit silicon carbide used previously to remove ink stain from the white pigmented epoxy. The pores appear white against the dark background, and the specimen is ready for microscopy or image analysis.

SUMMARY

Petrographic methods of sample preparation to analyze a cement or concrete specimen can be a useful tool for quality control and failure analysis. This paper describes various methods of specimen preparation for microscopy as well as for image analysis. A new technique to enhance contrast for image analysis in concrete and preparing ultra-thin sections is described, and the application of improved preparatioon equipment is emphasized.

REFERENCES

[1] Gouda, George R., "Effective Clinker Composition on Grindability," Cement and Concrete Research, vol. 9, 1979.

[2] Ahmed, W. U., "Preparation of Cement Clinkers for Microscopic Analysis," Proceedings of the Third International Conference on Cement Microscopy, Houston, Texas, pp. 1-9, International Cement Microscopy Association, 1981.

[3] Campbell, D. H. and Ahmed, W. U., "New Methods of Sample Preparation of Portland Cement Clinker and the Related Materials," Microstructural Science, vol 7, Elsevier, North Holland, Inc. New York, 1979, pp. 369-375.

[4] DeLisle, F. A., "Microscopic Analysis of Clinker and Cement." Cement Technology, May/June 1976, pp. 93-99.

[5] Ahmed, W. U., "Advances in Sample Preparation for Clinker and Concrete Microscopy", World Cement, vol 22, August 1993, pp. 8-12.

Donald H. Campbell[1]

A SUMMARY OF ONO'S METHOD FOR CEMENT QUALITY CONTROL WITH EMPHASIS ON BELITE COLOR

REFERENCE: Campbell, D. H., **"A Summary of Ono's Method for Cement Quality Control with Emphasis of Belite Color,"** Petrography of Cementitious Materials, ASTM STP 1215, Sharon M. DeHayes and David Stark, Eds., American Society for Testing and Materials, Philadelphia, 1994.

ABSTRACT

The use of Ono's Method for cement quality control is standard in plants of the Onoda Cement Company and other producers in Asia. Widespread use of the technique is found in cement plants of North and Central America. Alite birefringence, alite size, belite size, and belite color are reported to correlate with kiln conditions and hydraulic activity. The present paper is a preliminary evaluation of belite color, particularly as it relates to 28-day mortar-cube strength. Thirty samples of portland cement clinker or cement have been examined microscopically to determine the color of belite after extraction of the matrix phases (aluminate and ferrite) with a potassium hydroxide-sugar solution. A positively sloping regression line was found with a plot of percentages of clear plus pale yellow belite crystals (single and nested) and 28-day mortar cube strength; negatively sloping lines were found with plots of percentages of single and nested yellow and amber crystals, plus crystals with abundant dotlike impurities. Both single and nested crystals exhibit these relationships with 28-day strength.

Keywords: alite, belite, belite color, cement, clinker, cooling rate, microscopy, mortar strength, Ono Method, 28-day strength.

REFERENCE: Campbell, D.H., **A Summary of Ono's Method for Cement Quality Control with Emphasis on Belite Color,** *Petrography of Cementitious Materials*, ASTM STP 1215, S.M. DeHayes and D.S. Stark, Eds., American Society for Testing and Materials, Philadelphia.

---

[1]Senior Principal Petrographer, Petrographic Services, Construction Technology Laboratories, Inc., Skokie, Illinois, 60077 USA.

INTRODUCTION

The "Ono Method" for cement quality control, originated by Yoshio Ono of the Onoda Cement Company in Tokyo, was formally introduced to the western world in 1968 [1] and detailed in 1975 [2] at the Hawaiian seminar in which Ono taught a dozen North American cement microscopists the technique of interpreting kiln conditions and predicting the 28-day strength of an ASTM C-109 mortar cube. Such a revolutionary approach to cement quality control was, expectedly, met with some skepticism and even rejection. Interest and scientific curiosity, however, have continued and Ono's Method has become "standard practice" in many cement plants.

The late George Vanisko, microscopist at the Portland Cement Association (PCA), soon after attending the 1975 Hawaiian seminar, tested the Ono theory with plant and laboratory clinker and, in concert with others making similar tests, recommended to Nate Greening, then Director of Physical and Chemical Research at PCA, that the Ono Method had a sound observational and interpretive validity. The present author, still somewhat doubtful, was finally convinced when Vanisko demonstrated how the observations and interpretations based on reflected-light microscopy repeated those drawn from transmitted-light microscopy (the powder-mount method of Ono and the present writer's thin sections). Many of the same observations were made with each technique. Consequently, the PCA established a course in clinker and cement microscopy, now given three times per year with a recently introduced application of video tape. The Ono Method was the principal stimulus for the founding of the International Cement Microscopy Association (ICMA).*

Ono's theory of cement quality control was developed as part of his research for the doctoral degree at the University of Tokyo under the auspices of Onoda Cement Company during the early 1950s and summarized in his 1954 dissertation (in Japanese) [3]. Parts of his Ph.D. research are revealed in a series of later publications in English in which belite is given a relatively great emphasis and importance. Today, silicate microscopy and a prediction of 28-day mortar strength are utilized as statistical factors in computerized daily kiln-adjustment equations [4]. The rapidity with which a qualified microscopist can gather the required data (alite size and birefringence, belite size and color) has made Ono's method economical, providing an interpretation of clinker quality <u>before</u> it is ground into cement. Fuel energy costs make it desirable to optimize belite hydraulic reactivity because relatively low temperatures are required for belite crystallization, compared to alite.

A suitably proportioned and blended raw mix develops alite and belite crystals with relative ease. The rapid crystallization of alite forms a loosely linked framework of crystals which, for the most part, develop the "support structure" within which most of the other phases are scattered. Alite is always abundant, and in North America, usually large. That portion of the belite not converted to alite continues to grow at high temperature and, ideally, is well-scattered throughout the clinker. However, as a result of less-than-optimum feed grinding, nests of tightly packed belite remain and the nests can be considered as "hard knots" in an otherwise relatively uniform mesh of linked alite. Belite crystal size, morphology, color, and dispersal are major variables in many North American clinkers. Belite, the determinant of 28-day (and later) mortar strength, has an unusually complex microstructure of up to six sets of intersecting lamellae produced during polymorphic inversion within one crystal and the effects of exsolution (crystal unmixing). Belite requires specific attention with regard to cooling, a bit less important for alite, but not ignored. Granted there are many other very important factors in cement hydration (aluminate type, distribution, and abundance; cement fineness; ambient temperature; sulfate percentage and type;

---

*ICMA, 1206 Coventry Lane, Duncanville, Texas, 75137 USA.

compatibility with various admixtures, alite crystal chemistry, alkali level, crystal size, to name a few), but these factors are beyond the scope of this paper.

The purpose of this paper is to present briefly the rudiments of Ono's Theory and Method of cement quality control, with emphasis on the observation and recording of belite color. Belite color observations are compared mostly from two categories of commercial gray portland cement, (1) cements producing ASTM C-109 mortar cube strength over 48 MPa (7000 psi) at 28 days and (2) cements giving C-109 mortar strength of less than 38 MPa (5500 psi) at 28 days. A few white portland cements have also been studied. The analysis also includes the three standard reference clinkers offered by the National Institute of Standards and Technology, Gaithersburg, Maryland.

SUMMARY OF THE ONO METHOD

Table 1 lists the defining ranges and divisions of the four parameters (alite size, alite birefringence, belite size, and belite color) in relation to the kiln conditions and potential hydraulic activity. The divisions within each parameter are somewhat arbitrary, based on plant and laboratory data.

Table 2 lists the relation of these parameters to the 28-day mortar-cube strength as correlated by Ono using cements of normal fineness 350-400 (m^2/kg). The table numerically shows that belite color and alite birefringence have the most significant effects on 28-day mortar strength, other parameters kept constant. A description of Ono's Method and other applications of microscopy to cement and clinker are found in Campbell (1986) [7], from which this summary was taken.

Alite Size -- This parameter is a measure of kiln heating rate, the rate of temperature increase from approximately 1200 to 1450°C (2200 to 2650°F). Within this "transition zone" belite quickly changes to alite during a rapid incorporation of CaO. Thus from large belite comes coarsely crystalline alite. Ono has shown that "soaking" the clinker at high temperature (1450°C) for extended time (up to 20 hrs) increased crystal size to only 20-25 micrometers. In a uniformly operating kiln, with feed of constant fineness, relative changes of alite size may be interpreted to be the effect of heating (burning) rate. Average alite size (crystal length) is most accurately measured in clinker polished section, thus removing the effects of cement grinding. Crystal length means the longest dimension.

Alite Birefringence -- The numerical difference in indices of refraction of the fast and slow rays in an alite crystal is defined as the birefringence. Ono correlated alite birefringence with maximum temperature. Crystals with length-to-width ratio of approximately 2:1 and with relatively bright interference colors, are selected because in this crystallographic orientation maximum divergence of the two light rays occurs and maximum difference in refractive index can be measured. Because the alite crystal thickness is not usually precisely known and because of certain crystallographic reasons (optical directions, zoning, strain), the birefringence determined in powder mount is not the true birefringence and, therefore, should be termed "apparent birefringence." Treatment of cement or clinker with a KOH-sugar solution to remove the matrix greatly facilitates the determination of alite birefringence by setting free many crystals otherwise bound and obscured by matrix.

Belite Size -- Belite crystals not converted to alite at high temperatures (above 1400°C or 2550°F), are somewhat enlarged as they take in additional lime, silica, and impurities from the molten matrix and reach the alpha-polymorph stage. Belite must remain at high temperature for a sufficient time for additional crystal growth. Belite size (average crystal length) is most accurately measured with a clinker polished section, as is the abundance and distribution of belite nests and their crystal packing. As with alite, belite crystal length means its longest dimension.

Table 1 - BURNING CONDITION AND MICROSCOPICAL CHARACTER OF ALITE AND BELITE (ONO, 1981) [5]

Burning Condition	Hydraulic Activity			
	Excellent (+)	Good (vv)	Average (v)	Poor (-)
Heating Rate	Quick	-	-	Slow
Size of alite (μm)	15-20	20-30	(25) 30-40	40-60 (120)
Maximum Temperature	High	-	-	Low
Birefringence of alite	0.010-0.008	0.007-0.006	0.006-0.005	0.005-0.002
Burning Time	Long	-	-	Short
Size of belite (μm)	(20) 25-40 (60)	(15) 20-25	(10) 15-20	5-10
Cooling Rate	Quick	-	-	Slow
Color of belite	Clear (C)	Faint Yellow (FY)	Yellow (Y)	Amber (A)
Birefringence of belite	0.012	0.015	0.017	0.018
Content of alpha	Abundant (40%)	Medium (20%)	Few (10%)	Nil (0%)

Note: If MgO in clinker is higher than 1.8% or SO_3 above 1.25%, subtract 0.001 from your reading of birefringence. If MgO is less than 1.2%, add 0.001 to your reading.

TABLE 2 - EFFECT OF BURNING CONDITION ON 28-DAY STRENGTH, KG/CM²: EXCELLENT (+); GOOD (VV); AVERAGE (V); POOR (-); (ONO, 1980) [6]

Max. Temp		+				VV				V				-			
Burn Time		+	VV	V	-	+	VV	V	-	+	VV	V	-	+	VV	V	-
Rate of Burn	Cool																
+	+	470	465	460	455	445	440	440	435	425	420	415	410	405	400	395	390
	VV	445	445*	440*	435	425	420*	415*	415	405	400	395*	390	380	375	375	370
	V	425	420	415	415	405	400	395	390	380	375	375*	370	360	355	350	350
	-	405	400	395	390	380	380	375	370	360	355	350	350	340	335	330	325
VV	+	460	460	455	450	440	435	430	430	420	415	410	405	395	390	390	385
	VV	440	435*	430*	430	420	415*	410*	405	395	395	390*	385	375	370	365	365
	V	420	415	410	405	395	395	390	385	375	370	365*	365	355	350	345	340
	-	395	395	390	385	375	370	365	365	355	350	345	340	330	330	325	320
V	+	455	450	445	445	435	430	425	420	410	410	405	400	390	385	380	380
	VV	435*	430*	425	420	410*	410*	405*	400	390	385*	380*	380	370	365	360	355
	V	410	410*	405	400	390*	385*	385*	380	370	365*	360*	355	345	345	340	335
	-	390	385	385	380	370	365	360	355	345	345	340	335	325	320	315	315
-	+	450*	445	440	435	425	425	420	415	405	400	395	395	385	380	375	370
	VV	430*	425*	420	415	405*	400	400	395	385	380	375	370	360	360	355	350
	V	405	400	400	395	385*	380	375	370	360	360	355	350	340	335	330	330
	-	385	380	375	375	365	360	355	350	340	335	335	330	320	315	310	305

Note: Underline: Japanese cements in 1962-1963, 500-700 t/d kilns.

Asterisk (*): Japanese cements in 1975-1979, 3000-8000 t/d kilns; (kg/cm^2) (0.09807) = MPa; (kg/cm^2) (14.22) = psi.
psi (0.006894) = MPa

Belite Color -- Belite color is determined during cooling from maximum temperature to approximately 1000°C (1830°F). Slowly cooled belite has time to exsolve some of its high-temperature entrapped impurities, many of which accumulate beside and within lamellae, producing a pale yellow to amber coloration. Exsolution during cooling and polymorph inversion should, therefore, be minimized. The determination of belite color in a powder mount requires one's ability to make judgments on crystal colors occurring on the borderlines between the major categories. Clear crystals indicate rapid cooling and relative preservation of the alpha polymorph; amber colors suggest slow cooling and relative decrease in the alpha polymorph. Ono's interpretation of slow cooling also includes clear crystals with abundant dot-like impurities from exsolution. Belite lamellar extensions into the matrix as the crystals continue to grow during slow cooling is readily observed in polished sections. Jagged crystal surfaces are commonly seen in KOH-sugar treated powders. Rapid cooling not only retards exsolution but forces matrix crystallization, thereby forming a solid barrier that impedes belite lamellar extension.

"ADJUSTMENTS" TO A 28-DAY MORTAR-STRENGTH PREDICTION

Although belite cooling rate is a major factor in developing 28-day mortar strength, other cement factors also have significant effects, namely, the cement fineness and particle size distribution, belite percentage, and degree of belite nest development, among other factors. The extensive evaluation of these factors is beyond the scope of the present paper, however, a few comments are necessary.

Very finely ground cements (surface areas of greater than 550 m^2/kg) typically show a relatively high mortar strength at 28 days. One might assume that some of the undesirable effects of slowly cooled belite or belite nests can be in part mitigated by fine grinding, thus increasing the hydraulically active surface areas of all clinker phases. A sufficient number of these finely ground cements have not been studied by the writer to warrant statistically-based correlations regarding fineness, belite nests, and belite color, assuming such correlations exist. Cement science is in great need of statistically-based systematic microscopy.

Belite nests have a major control on late strength gain for two principal reasons: (1) Many belite nests in which the crystals are tightly packed (originating from coarse silica-rich particles) tend to remain as such during clinker grinding, thereby seriously reducing the effective surface area of belite exposed for reaction with mix water and curing moisture. Even though the belite crystals may be clear, their availability to react with water is lessened when crystals are combined in nests. Consequently, a "correction factor" based on the percentage and type of belite nests could be applied to the prediction of 28-day mortar strength given in Ono's table, or as Prout astutely recommends, the table value could be used as an "index number" instead of actual strength. For some plants, other characteristics of the cement (e.g. alite size, birefringence, crystal chemistry, fine grinding, etc.) appear to override the negative effects of the nests.

Examination of powder mounts of the greater-than-45 µm fraction of a cement (as received) clearly indicates the relative abundance of belite nests from sample to sample; the same conclusion can be, perhaps, more accurately drawn from polished sections. A relatively large number of belite nests might be tolerated in a clinker if, in the same clinker, belite is also well scattered as solitary crystals. Theoretically, belite crystals should be clear, sufficiently large, and uniformly distributed in the clinker, thus promoting an optimum hydration for late strength.

BELITE COLOR OBSERVATION

In order to evaluate the importance of belite color, samples of many cements representing relatively high and low 28-day strengths (greater than 7000 and less than 5500 psi, respectively) were sieved to produce the 45- to 75-µm fraction, which was then treated with the KOH-sugar solution to remove the aluminate and ferrite. In practice, the KOH-sugar treatment actually produces many silicate crystals less than 45 µm because the matrix previously holding the crystals together in the sieve fraction is dissolved. The treatment, however, makes the belite color an easily observable characteristic in powder mount. Use of the 45- to 75-µm fraction should not be considered mandatory. Statistically treated comparisons of silicate characteristics in various sieve functions (after extraction) would be an interesting exercise and might have important applications, however, that research has not been published. Some reasons for using the 45- to 75-µm fraction are given in Discussion and below. Observational problems, however, can clearly skew counting data and complicate the interpretation of cooling rate:

1. It is well known in mineralogy that many minerals may be strongly colored in hand specimen but colorless when viewed as small particles under the petrographic microscope (e.g., microcline, fluorite, quartz, dolomite, calcite, etc.). In addition, with a smaller particle size, like, the 10- to 15-µm fraction of the cement or crushed clinker, one might find a tendency to judge a crystal fragment color to be clear when actually the color of the original whole crystal was pale yellow. The natural color of belite crystals in plane-polarized light, like interference colors in cross-polarized light, are functions of crystal thickness. Large wedge-shaped fragments of single belite crystals tend to be clearer and always have lower interference colors on the thin side of the particle than on the thick side. In the present research the relatively thick side was chosen as the color indicator. However, for nests that survive the KOH-sugar treatment, the problem is not as easily resolved. Small round belite crystals are commonly clear (in anyone's definition) but, when stacked in a nest like a handful of marbles, the clear color may appear tinted because of tiny masses of ferrite or other phases buried deeply in the nest between belite crystals and thus protected from the KOH-sugar solution, or due to Becke line effects. When counting nested crystals along the line of traverse, one must consider each crystal and make the best judgment possible.

   Tiny belite crystals formed as a surficial decomposition of the alite during slow cooling have always been clear in the writer's observations; secondary belite (formed out of the matrix during cooling) is also clear, perhaps because these crystals have little to exsolve. Research is needed on this point. Amoeboid crystals are rarely colored; dendritic crystals have never been observed to be colored in the writer's experience. Ono's belite color interpretation seems most applicable to "properly" formed crystals.

2. "Boundary-line colors" refers to colors falling between the major categories of clear, pale yellow, yellow, and amber. In other words, pale yellow to one observer is obviously yellow to another, even when each person is being as objective as possible. This problem is not unique to cement microscopists! What is needed? A photographic belite color standard in the form of 35-mm slides (or prints) as a companion to the present paper is presently being prepared by the Portland Cement Association for general use by cement microscopists.

3. The "color" one observes is not altogether a product of exsolution, but also of crystal microstructure. Changes in light-ray velocity and direction when passing through materials of nearly identical refractive index (belite lamellae, for example) commonly produce faint bands of color or light intensity as the waves interfere. These effects can be linear, that is occurring along the boundaries of adjacent lamellae, easily visible in thin wedge fragments or more complexly formed by intersecting lamellae. The Becke line itself may have a yellowish color in liquid with a refractive index of 1.715 or Hyrax* (1.70), and care should

*Custom R&D, 8500 Mt. Vernon Road, Auburn, CA 95603, (916) 885-3341.

be taken not to confuse it with crystal color; the level of focus should be such that the Becke line is just outside the belite crystal for color determination. The Becke line around alite can give the crystal a yellowish color as the focus is raised or lowered.

4. Certain cations such as chromium, manganese, and, perhaps, aluminum, and others from refractory bricks can produce a range of green and yellow variations in belite. These causes of coloration do not appear related to cooling rate, however, much research remains The relation of crystal chemistry and color to compositions of waste solvent kiln fuels is little known. The absence of colored belite in slowly cooled white cement clinker suggests that iron is the primary colorant in other portland cements.

EXTRACTION TECHNIQUE AND COUNTING PROCEDURE

Powder mounts of the silicate residue resulting from a KOH-sugar extraction procedure were prepared for transmitted polarized-light examination using a Nikon Optiphot microscope. The procedure for extracting the aluminates and ferrites and studying the silicate residue is as follows:

1. Sieve the crushed clinker or cement and retain the 45 to 75 μm (325-to-200 mesh) fraction. Use isopropyl alcohol from a spray bottle to facilitate sieving and to produce a clean powder.

2. Rinse the 45-75 μm fraction with a thin stream of isopropyl alcohol into a plastic weigh boat or polyethylene container. Decant or absorb the excess alcohol with paper towels and dry the powder.

3. Transfer the dry powder into a small labeled vial or other suitable container.

4. Place a 125-mm diameter watch glass or small beaker on a ceramic triangle on a hot plate set so that the liquid temperature will be approximately 80°C (176°F).

5. Place a few milligrams of the dry powder in the watch glass and add 5-10 mL of the KOH-sugar solution (5 grams table sugar + 10 grams KOH + 20 mL deionized water). <u>This solution is extremely caustic</u> and, therefore, proper laboratory safety precautions such as protective eye wear and rubber gloves are strongly recommended in steps 5 and 6. Stir the mixture occasionally with a clean glass rod while warming for approximately 10 minutes. Remove the watch glass from the hot plate. Withdraw the hot liquid with paper towels or decant.

6. Rinse the wet residue in the watch glass with a stream of deionized water from a polyethylene squeeze bottle, decanting the water or withdrawing it with paper towels. Do this at least three times within a 3-minute interval. Do not be concerned with traces of a precipitate that may form or tiny crystals growing on the clinker particles.

7. Flood the residue in the watch glass with acetone and decant or withdraw the liquid with paper towels. Do this three times over a time interval of approximately three minutes.

Allow the residue to dry and transfer into a labeled vial. Slick paper or a small plastic weight boat and an artists paint brush facilitate the transfer.

8. Prepare a powder mount, using refractive index liquid (n=1.715, approximately), for examination in transmitted polarized light. Hyrax, a synthetic resin, may also be used to make a permanent powder mount (see Campbell, 1986) [7]. The microscope is equipped with a "daylight blue" filter for routine observations and photomicrography. The same powder mount is also used to determine the apparent birefringence of alite.

9. Counting Procedure and Data Reduction -- Using a manual mechanical stage, make regular traverses at 300 µm intervals across a slide area of approximately 25 mm^2. Only crystals occurring on the line are tallied. The two major categories for the present study were "Single Crystals" and "Nests," and within each of which the crystal color was recorded as "Clear," "Pale Yellow," "Yellow," or "Amber;" two additional categories were applied where needed, "Crystals with Abundant Dotlike Inclusions" and "Green Crystals." A nest was defined as two or more crystals in a single particle; no determination was made as to tightness of the crystal packing in the nests. Consequently, in a nest of 15 small crystals, only those 5 crystals on the line, for example, would be tallied. Belite inclusions in alite are not recorded. Average number of crystals tallied per slide in the present study is 238.

   Each powder mount in the present study was number coded so that prior knowledge of strength characteristics would not bias the data, and not until all data were gathered were statistical calculations made. White cement data were not used in statistical calculations because of the general lack of belite color due to iron scarcity in the raw feed. Following the procedure recommended by Prout [8], belite colors can be recorded numerically using a continuous scale (Clear = 4.0, Faint Yellow = 3.0, Yellow = 2.0, and Amber = 1.0), thus crystal colors might be recorded as 3.8, 2.2, 1.5, etc. Clear crystals with abundant dotlike inclusions are said by Ono to result from exsolution during slow cooling; a numerical ranking of 2 was arbitrarily applied to these crystals by the present writer. Linear regression equations and correlation coefficients were calculated for selected groups of data.

10. Polished sections, considered indispensable and part of routine work in clinker and cement microscopy, were made. The sections were prepared by placing thick epoxy-cement slurries of the 45- to 75-µm fractions separately on a glass microscopic slide, heaping additional dry sample onto each slurry, and allowing to harden at room temperature. Three to four slurries were placed on each slide. The tops of the hardened slurries were then ground and polished in the normal manner (see Campbell) [7] using a thin-section slide holder and horizontal rotary wheels or other devices. Thickness of the resulting polished section is roughly 0.25 mm. These preparations are not thin sections because the encapsulated particles have been sectioned only on one side (thin sections have planar particle cross sections on both top and bottom surfaces and the sections are of nearly uniform thickness). The polished sections were used to make a qualitative correlation between belite colors (from a treated-powder mount) and lamellar extensions into the matrix, the relation apparently needing additional study. A simple rapid procedure for preparation of multiple polished cement or clinker thin sections is needed; it might render the KOH-sugar technique unnecessary and provide silicate-matrix and matrix data as a bonus.

DISCUSSION

Some readers might question the use of the 45- to 75-μm crushed clinker or cement fraction in the determination of the kiln-condition parameters by Ono's method, saying that such a particle-size range does not fairly represent the material. Granted, a chemical analysis of the greater-than-45-μm fraction of many cements and crushed clinkers may have a higher silica percentage than the finer fraction because of the tendency during initial grinding to concentrate belite nests. Therefore, the coarse fraction probably does not represent the bulk chemical analysis. But a chemical analysis is not microscopy. In the Ono method we are concerned primarily with certain microscopical aspects of silicate crystals. A standard size range recommended in optical mineralogy texts is 74 to 149 μm, for mineralogical analysis. For unusually finely ground cements or cements produced with virtually all particles less than 45 μm, the size range discussed in this paper is probably not practical.

Several reasons may be given for use of the 45- to 75-μm fraction:

1. Because the KOH-sugar solution separates many of the silicate crystals within the original 45- to 75-μm particle, crystals down to 5 or 10 μm, or smaller, are abundant in the powder mount. Thus what was originally a 45- to 74-μm fraction actually becomes a fraction with a much lower, probably submicrometer, minimum size, and therefore, small crystals are well represented in the powder mount. One should be wary of crystals too small to produce a usable interference color or natural color.

2. Whole crystals are quickly found for examination. In addition, the silicate optical properties are relatively easily observed, especially alite interference color and the natural color of belite.

3. Crystal sizes, as the present writer utilizes Ono theory, are measured in polished sections, not powder mount. A polished section, containing whole or fragmented clinkers, is clearly a better way of measuring the average (most frequently occurring) crystal size than a finely crushed sample which typically contains many fragments of larger crystals. However, if by examination of a 45- to 75-μm fraction in a powder mount, one sees only fragments of large silicate crystals, then the average original crystal size is clearly quite large, and consequently, according to Ono theory, the burning rate (alite size) was slow and the burning time (belite size) was long. In this case, actual measurement of crystal size in polished section is unnecessary for the interpretation.

4. An alcohol wash while sieving produces a relatively clean assemblage of particles, free of micrometer-size "dust" that typically obscures examination.

The microscopical data derived from the KOH-sugar treated 45- to 75-μm fraction are believed to be interpretable in terms of the Ono Method, but this size range should not be taken as a requirement. Preliminary data suggest no significant difference from the less-than-45-μm fraction. If these interpretations can be shown to be erroneous, then other techniques should be investigated. An argument could be made for standardizing a size fraction to be used in the Ono-Method powder mount.

BELITE COLOR VARIATION AND 28-DAY STRENGTH

Correlation coefficients derived from data representing 23 samples given in Table 3 indicate only gross trends and weak correlation of belite color and 28-day strength, suggesting, as expected, that other complicating factors, namely the alite and belite size and alite birefringence,

	Single Crystals (%)					Nested Crystals (%)					28-day strength (psi)[f]
	Clear A	Pale Yellow B	Yellow C	Amber D	Dotlike Impurities[e] E	Clear F	Pale Yellow G	Yellow H	Amber I	Dotlike Impurities[e] J	
1	7.8	8.6	6.0	0	4.7	12.1	27.6	26.7	4.3	2.2	5217
2	29.2	33.0	6.1	1.4	3.8	11.3	13.2	1.9	0	0	5127
3	4.0	8.4	23.5	4.0	0.5	4.0	17.9	28.7	7.2	1.6	4367
4	12.6	21.2	26.4	3.0	2.6	5.2	12.6	10	5.6	0.9	6800
5[a]	53.0	1.7	0.4	0	4.8	36.5	0.4	0	0	3.0	8720
6[a]	43.5	7.0	0.4	0	5.2	36.5	5.7	0	0.9	0.9	8720
7	31.9	21.1	7.8	0	2.9	25.0	7.4	2.0	1.0	1.0	5000
8	1.6	3.1	4.7	0.8	3.1	25.5	17.3	32.2	7.0	4.7	7230
9	Belite only occurs as secondary alteration on surfaces of alite; no statistical data										4812
10	1.4	1.4	7.5	6.6	1.4	16.4	12.2	23.9	29.1	0	4800
11	2.7	10.7	32.1	1.8	0.5	1.8	10.7	23.2	15.2	1.3	5507
12	5.4	9.2	2.9	1.7	3.8	8.8	35.0	26.7	0.8	5.8	6810
13[a]	62.0	7.9	0.5	0	5.1	17.3	4.7	0	0	2.3	8100
14	4.6	19.2	17.5	12.1	7.9	3.3	10.8	17.9	5.8	0.8	7100
15	7.8	6.6	2.7	1.2	1.2	40.2	21.9	8.2	4.7	5.5	7220
16	43.7	24.9	10.3	3.8	5.2	5.2	5.6	1.4	0	0	7080
17	12.8	20.7	15.0	3.4	9.4	6.8	13.1	10.5	4.9	3.4	7100
18	3.0	16.3	15.9	5.2	2.1	10.3	26.2	18.0	3.0	0	6970
19	2.9	18.0	45.6	3.8	2.1	1.7	5.0	17.2	2.5	1.3	5507
20	0.7	3.2	11.9	7.7	5.9	14.3	11.5	22.7	17.8	4.2	4700
21	28.7	28.7	8.1	3.3	3.8	3.8	4.8	5.7	4.8	8.1	5507
22	22.4	21.1	5.4	0.9	4.0	7.2	26.9	8.1	0.9	3.1	6810
23	14.7	9.1	8.2	1.7	5.2	3.9	26.7	20.7	5.2	4.7	5120
24	1.3	2.9	6.7	13.3	31.3	0	5.4	9.6	20.8	8.8	5900
25	28.6	9.9	2.4	0.8	3.6	20.6	10.7	11.1	4.4	7.9	7040
26[b]	18.5	9.1	0.4	0	1.1	23.4	38.9	1.9	0.4	6.4	NA
27[c]	10.4	1.6	0	0	1.2	38.6	41.0	4.8	0	2.4	NA
28[d]	33.1	35.6	12.7	3.7	10.2	0	2.1	2.9	0	0	NA
29[a]	38.8	0	0	0	1.0	60.2	0	0	0	0	8000
30	4.8	17.0	14.8	6.5	7.4	3.5	8.7	19.6	10.9	7.0	6600

Linear Regression Equations and Correlation Coefficeints:

A+B $y = 5900.9 + 6.2797x$, $R^2 = 0.016$

C+D+E $y = 6241.6 - 7.9680x$, $R^2 = 0.012$

F+G $y = 5703.9 + 14.279x$, $R^2 = 0.043$

H+I+J $y = 6558.6 - 18.769x$, $R^2 = 0.074$

A+B+F+G $y = 5462.0 + 11.608x$, $R^2 = 0.064$

[a]White cements; not considered statistically.

[b][c][d]NIST Standard Reference Materials, 8486, 8487, 8488, respectively. Gaithersburg, Maryland,

[e]Generally clear crystals.

[f]ASTM C109 mortars, made and tested at CTL.

plus the many variables of cement mentioned previously, exert strong controls on the performance of cement. Therefore, it was not surprising that "tight," clearly linear, graphic relationships between belite color and 28-day strength were not defined. However, positively sloping linear regression lines are evident with clear and pale yellow crystals, occurring singly and also in nests. Negatively sloping lines result with yellow and amber crystals (including those with abundant dotlike impurities) likewise occurring singly and in nests. Ono's theory of strength improvement with increasing percentages of clear and pale yellow belite appears sound; conversely, yellow and amber crystals, reflecting crystal exsolution during slow cooling, produces a cement with less-than-optimum performance.

CONCLUSIONS

The samples chosen for analysis represent many different production methods (preheater versus wet process) and a wide variety of raw materials, thus different crystal chemistries. Where samples from only one kiln are analyzed, one might expect less influence of these major complicating factors and, consequently, the belite color correlation with 28-day strength may be much better. The present research indicates weak correlations with belite color and 28-day strength of mortar, however, the analysis is far from definitive. Prout (personal communication) has recommended investigation of belite color in relation to strength gain from 7 to 28 days. Such detailed systematic statistical research with a sufficient number of samples and crystal counts remains largely unpublished. In other words, a within-plant, single kiln analysis, or better carefully controlled laboratory studies (like those done by Ono and colleagues in Japan) may reveal significantly higher correlations. Suffice it to say, our efforts in systematic statistical microscopy for modern kiln optimization, as well as many other aspects of our industry, are deplorably scarce, less than rigorous, and dishearteningly inadequate. I firmly believe that microscopical analysis is one of the most economical, most accurate, most revealing techniques of quality control and diagnosis of problems. Without it we are visually impaired.

ACKNOWLEDGMENTS

The writer is grateful to the Portland Cement Association for funding of this project (PCA Project Index No. 92-8). The text portion of this report (PCA R&D Serial No. 1962) will be included in a 35-mm slide set (SS-403) to be offered through PCA for sale in 1993. I am also grateful for the statistical analyses done by Ms. Wilma Dziedzic, Manager, Physical Testing Services, CTL; the calculations were done with a MAC II_{si} and the Cricket Graph program. The contents of this report reflect the views of the author who is responsible for the facts and accuracy of the data presented. The contents do not necessarily reflect the views of the Portland Cement Association.

REFERENCES

[1] Yamaguchi, G., and Ono, Y., "Microscopic Studies on the Textures of Belite in Portland Cement Clinker," *Reviews*, 16th General Meeting of the Cement Association of Japan, 1962, pp. 32-34.

[2] Ono, Y., "Microscopic Analysis of Clinker," Onoda Cement Co., Central Research Laboratory, 1973/12/15 and 1975/6/22. Paper supplied to students at Hawaiian seminar in 1975.

[3] Ono. Y., "Study of Belite in Clinker," Ph.D. Dissertation, Tokyo University, 1963, 284 pp.

[4] Nakamura, G., Aizawa, T., and Nakase, K., "Optimization of Cement Manufacturing Process," Journal Research, Onoda Cement Co., Vol. 40, 1988, pp. 29-40.

[5] Ono. Y., "Microscopical Observation of Clinker for the Estimation of Burning Condition, Grindability, and Hydraulic Activity," *Proceedings of the Third International Conference on Cement Microscopy*, International Cement Microscopy Association, Houston, Texas, 1981, pp. 198-210.

[6] Ono, Y., "Microscopical Estimation of Burning Condition and Quality of Clinker," *Seventh International Congress on the Chemistry of Cement*, Paris, Vol. 2, Theme I, 1980, pp. 206-211.

[7] Campbell, D.H., *Microscopical Examination and Interpretation of Portland Cement and Clinker*, SP030T, Portland Cement Association, Skokie, Illinois, 60077, USA, 1986, 128 pp.

[8] Prout, J., "Summary of Interpretation of Data from the Ono Test," Portland Cement Association, Shortcourse in Cement and Clinker Microscopy, Notebook, 1980, 4 p.

Shondeep L. Sarkar,[1] and Basma Samet[2]

THE MICROSTRUCTURAL APPROACH TO SOLVING CLINKER-RELATED PROBLEMS

REFERENCE: Sarkar, S. L. and Samet, B., **"The Microstructural Approach to Solving Clinker-Related Problems,"** Petrography of Cementitious Materials, ASTM STP 1215, Sharon M. DeHayes and David Stark, Eds., American Society for Testing and Materials, Philadelphia, 1994.

ABSTRACT: The paper discusses in detail the principles underlying today's common microstructural investigative techniques and their capabilities in diagnosing problems associated with clinker manufacture and their properties.

Several characterization techniques are available to the cement scientist/technologist. These can be divided into two broad categories, namely direct observation techniques: optical microscopy and scanning electron microscopy (SEM). An additional tool used in conjunction with SEM is the energy dispersive X-ray analyzer (EDXA) unit which enables chemical analysis to be performed simultaneously. Indirect interpretative technique mainly comprises X-ray diffraction (XRD).

While all these techniques offer useful information and some definite answers concerning raw materials, thermal history of clinkerization, and resulting clinker properties, sample preparation method is of prime importance in order to derive accurate data. Each technique requires a specific sample preparation process. These are described at some length.

A number of case histories of microstructural and microanalytical characterization of clinkers are also presented in order to highlight their importance in solving clinker-related problems.

KEYWORDS: Clinker, microstructure, raw meal, kiln, optical microscopy,SEM/EDXA, XRD, alkali sulfate, C_3S, C_2S, C_3A, C_4AF.

Since portland cement was patented by J. Aspdin in 1824, it has won a firm position in modern society as a cheap, reliable basic construction material. Although cement production on a global scale today exceeds 1151 million tons per annum [1], the

[1]Research Professor, Department of Civil Engineering, Université de Sherbrooke, Quebec,Canada, J1K 2R1

[2]Graduate Student, Department of Civil Engineering, Université de Sherbrooke, Sherbrooke, Quebec, Canada, J1K 2R1

process is not totally free from problems, specially in terms of raw material selection, composition optimization, selection of proper clinkering conditions and raw meal grinding.

Clinker microstructure is directly related to inhomogeneities in the raw meal, and clinkering parameters such as burning, heating rate and cooling. Clinker mineral examination rather than its bulk chemical analysis therefore helps better understand the clinker thermal history. The quality of clinker produced can be determined from judicious microstructural investigation using optical microscopy, SEM-EDXA, XRD techniques (with supportive chemical analysis).

The hydration phenomenon, and therefore, related properties such as rheology (setting, grout flowability, concrete slump loss, etc.) and the strength development depend on interrelated factors such as clinker phase mineralogy, polymorphic forms of principal phases, and minor substituting ions in the structure [2]. In addition, minor components such as alkali sulfates, free lime, magnesia, etc. are also known to play an important role.

In preparation for discussing the importance of clinker microstructure (that is presented in the form of a series of case studies), the following section presents the principles underlying today's common investigative techniques that are used by cement technologists and researchers to solve problems related to clinker production, process technology and raw materials. The capabilities of these techniques are also discussed.

TECHNIQUES

Several microstructural characterization techniques are available for studying the clinker microstructure. These can be divided into two broad categories, as follows:

(a) Direct observation techniques: optical microscopy and scanning electron microscopy.

(b) Indirect interpretative techniques: X-ray diffraction (XRD) analysis, bulk chemical analysis by X-ray fluorescence (XRF), and energy dispersive X-ray analysis (EDXA).

Optical microscopy

Two petrographic methods of examination are available: (a) transmitted light, and (b) reflected light microscopy. The use of technique (a) among cement scientists dates back as early as 1897, when Tornebohm observed that clinker was composed of 'alit', 'belit', 'celit' and 'felit'.

The systematization of reflected-light microscopy for investigating clinker microstructure was first carried out by Insley [3] in 1940. Ever since then the technique has been widely used by cement technologists to evaluate clinker quality. As most of the phases exhibit nearly the same reflectivity even between crossed-polars, Ellson and Weymouth [4] produced a compendium of chemicals which may be used as etchants to differentiate phases in a cement clinker. The use of an etchant brings about preferential colours among various compounds, which enables the microscopist to clearly distinguish the phases. Campbell's [5] more recent concise list contains the commonly used etchants for clinker microscopy.

Even with the advent of several modern experimental techniques and analytical tools (such as SEM/EDXA, back scattered electron imaging, nuclear magnetic resonance, Ramanscopy, etc.) in the field of cement technology, optical microscopy still remains a useful tool for phase identification and systematic study of crystalline compounds in cement clinker. As a matter of fact, optical microscopy offers a relatively rapid and inexpensive means of clinker investigation.

Scanning electron microscopy

SEM is well entrenched as an electron column (industrial) instrument of the present day; therefore, it needs no detailed description. Nevertheless, it must be noted that the primary use of the SEM lies in its direct observation of surface topography from the detection of secondary reflected electrons. Low (x 20) to very high (x 10^5) magnification can be selected with relative ease, fractured and polished clinker sections viewed, and all with a depth of field much greater than that of an optical microscope, creating a 3-D effect. An example is provided in Figure 1. In most cases SEM imagery is easy to interpret, especially in combination with EDX analysis [6].

Over the last decade or two, EDXA has gained wide recognition among cement scientists as a highly versatile microanalytical tool. The rapidity with which complete interpretation can be performed using EDXA cannot be overlooked. Skalny et al. [7] in using this technique to study industrial clinkers, contend that it can be successfully applied (in combination with optical microscopy and XRD) in the identification of major as well as minor components of cement clinker. They found this method particularly useful for studying interstitial phases of different compositions, whereas the authors frequently use SEM/EDXA for obtaining confirmatory proof of alkali sulfate compositions in clinker (Figure 2), and for the identification of unusual minerals [8] that are rare occurrences as shown in Figure 3.

Most researchers, however, comment on the adverse effect of lateral beam spread on penetrating the specimen [9]. EDX analysis cannot be restricted to an arbitrarily small volume; spurious contributions from neighbouring or underlying grains can indeed jeopardize accurate analysis. Lowering the beam current/voltage may narrow this spread, but it reduces the sensitivity of the system. So, users have adapted themselves to a compensation point. Although attempts have been made to quantify this beam divergence, uncertainty still prevails for a composite material such as cement.

With modern ultrathin and windowless EDX detectors it is possible to detect carbon and oxygen, but only in qualitative terms. Nevertheless, this capability does help the cement microscopist in distinguishing carbonate/oxide/hydroxide minerals with similar morphology [6].

Application of atomic number, fluorescence and absorption (ZAF) corrections prove very effective in determining the accurate chemical composition of cementious phases, particularly when studying the role of minor components in cement. This is illustrated in Table 1, where one observes that most of the alkalies are incorporated in C_3A* and C_2S phases, Al and Fe are preferentially substituted in C_3S and C_2S lattices, and Mg in the interstitial phases. The ZAF software is now available with most EDX systems; its optimal performance dictates the use of flat polished specimens, though at the cost of phase delineation. The authors, however, have established that a very light etching with salicyclic acid + methanol is adequate for phase demarcation (even in the secondary electron mode), without any loss of accuracy in quantitative results. To overcome this phase recognition problem, backscattered electron imaging (BEI) is often used [10]. This technique utilizes another type of electrons that are emitted from the specimen in proportion to atomic number.

---

* Cement technologists' notations:
$C = CaO$, $S = SiO_2$, $A = Al_2O_3$, $F = Fe_2O_3$, $\overline{S} = SO_3$, $K= K_2O$, $N = Na_2O$

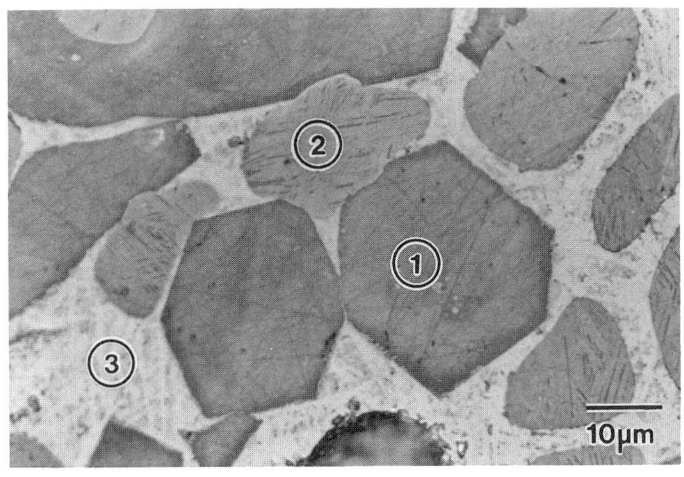

FIG. 1a.--The interior of a clinker, viewed under reflected light optical microscope. 1 - C_3S, 2 = C_2S, 3 = Interstitial matrix (C_3A + C_4AF)

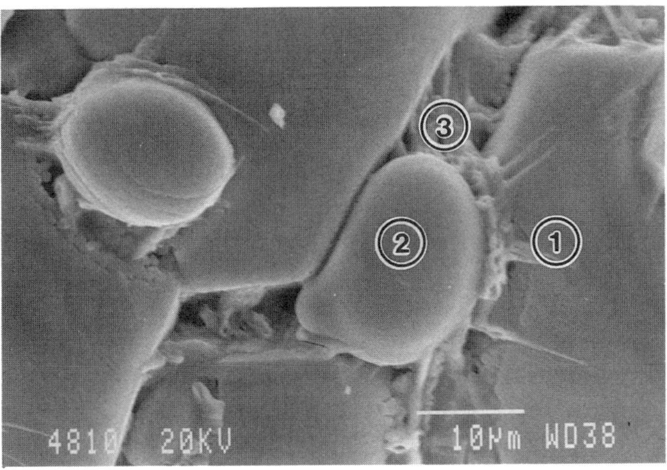

FIG. 1b.--A clinker when viewed under SEM, showing much greater depth of field and higher resolution. 1 = C_3S, 2 = C_2S, 3 = Interstitial matrix (C_3A + C_4AF)

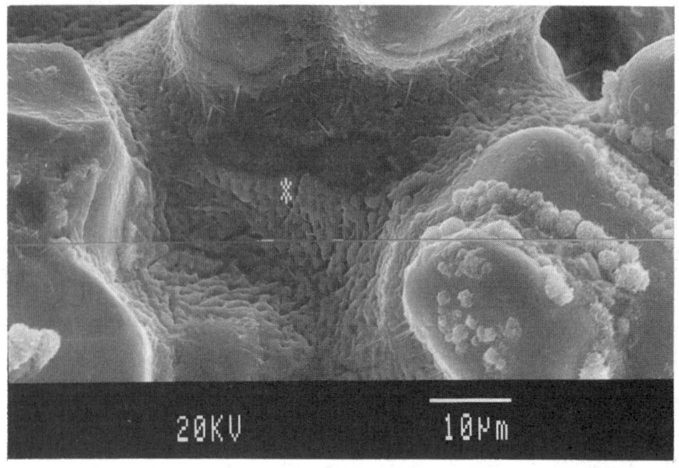

(a)

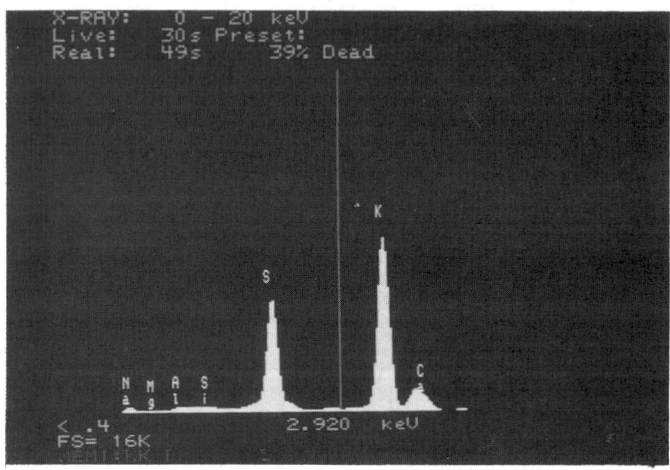

(b)

FIG. 2.--(a) Arcanite (K_2SO_4) in a clinker. (b) EDX spectrum of arcanite

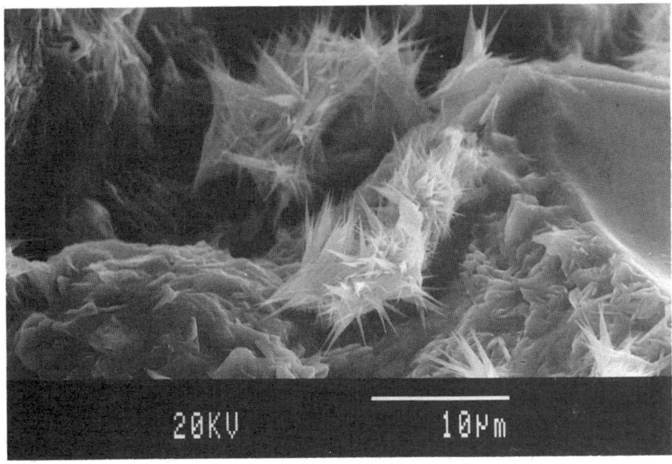

(a)

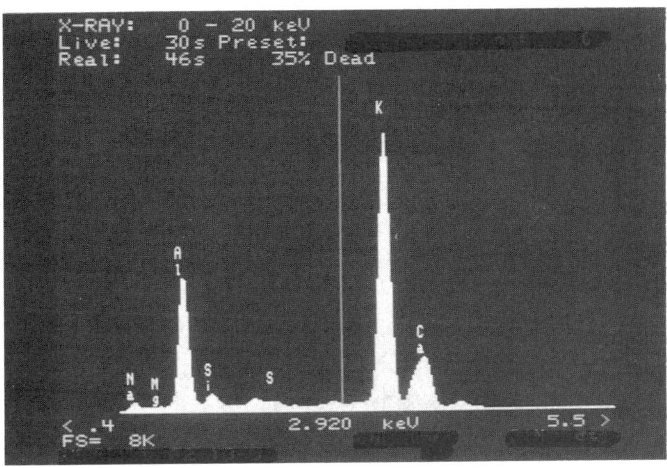

(b)

FIG. 3.--(a) Mixed crystals of K_2O and Al_2O_3 [8] also identified by the authors, in a clinker rich in alkalies. (b) EDX spectrum of these crystals

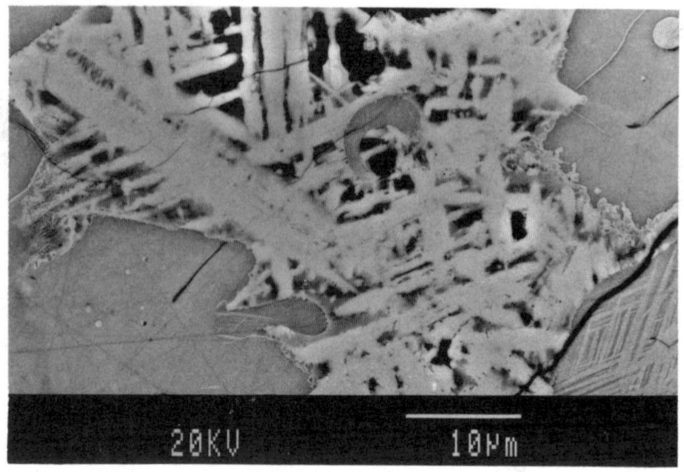

(a)

(b)

FIG.4.--Backscattered electron image of (a) coarsely crystalline interstitial matrix and (b) fine, glassy matrix

TABLE 1-- Quantitative analytical results of major and minor components in clinker, performed using EDXA-ZAF corrections

Phase	Ca	Si	Na	Mg	Al	S	K	Fe
C_3S	48.7	11.9	0.1	1.2	0.5	0.1	0.1	0.4
C_2S	43.5	15.5	0.2	0.5	0.8	0.4	0.4	0.5
C_3A	40.8	3.5	0.2	1.1	15.4	0.1	0.4	4.4
C_4AF	35.3	2.9	0.1	1.4	10.7	0.2	0.1	15.1

Greater utility for BEI can be found in the distinction of interstitial matrix phase morphology. Whether interstitial phases are coarsely crystalline or glassy can be readily determined from BEI (Figure 4). It is well known that the cooling rate strongly influences the crystallization of the liquid phase comprising C_3A and C_4AF; coarsely crystalline interstitial matrix in a clinker indicates slow cooling. The advent of a digital image processor attachment to EDXA, enables the operator to 'treat' these images digitally for clarity, feature enhancement, and phase delineation [11]. An example is shown in Figure 5.

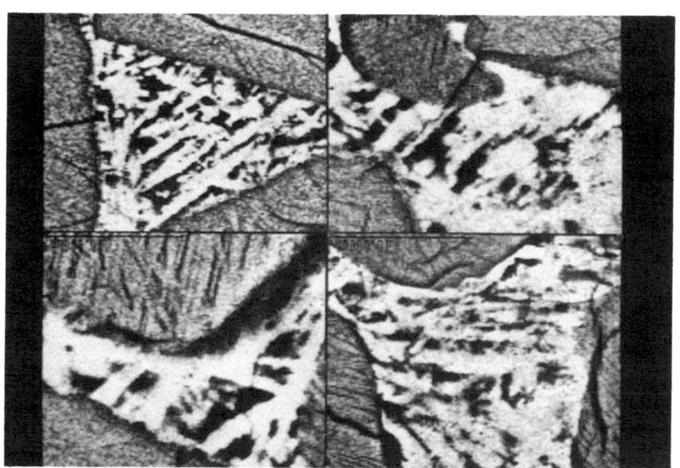

FIG. 5--Interstitial matrix of four clinkers after image treatment

X-ray mapping of selective elements is particularly advantageous as a supplementary tool for locating the actual distribution of minor elements in a clinker [12]. For example, K may be incorporated in the belite lattice, as well as in C_3A, or it can occur as $K_2SO_4/(NaK)_2\ SO_4/K_2Ca_2(SO_4)_3/K_2Ca(SO_4)_2 \cdot H_2O$ crystals. Similarly, Mg can be found independently as periclase (MgO) crystals, or it may be located in the aluminoferrite phase in the interstitial matrix, or substituted in C_3S grains. An illustration of clinker elemental X-ray mapping is provided in Figure 6. One observes in these maps that Mg occurs as periclase (MgO) crystals, and K and S as arcanite (K_2SO_4).

X-ray diffraction

The advantage of x-ray diffraction method as an interpretative tool for cement analysis lies in its ability to provide, in principle, accurate means of phase identification, rather than arriving at the composition by deductive calculations from chemical analysis, i.e., the classical Bogue method [13] and by optical counting methods. Although the Bogue principle provides valuable information about compound compositions, its assumptions of high temperature phase equilibrium, inability to delineate the polymorphic forms of the same compound, and the unaccounted minor components in portland cement that are of some significance, result in serious limitations and inaccuracies not found in the XRD method, even at the qualitative level.

For example, the phase composition of two clinkers from Bogue calculation were found to be very similar, as follows:

	C_3S	C_2S	C_3A	C_4AF
Clinker A	56.0	22.9	6.4	10.8
Clinker B	56.2	21.5	6.8	10.5

However, from XRD analysis, shown in Figure 7 it is evident that the total silicate phases in clinker B is distinctly higher, as is the C_4AF phase in clinker A. Differences in polymorphic forms of C_3A are also recognizable. Clinker B contains arcanite (K_2SO_4), whereas a trace amount of calcite is present in clinker A.

Proper interpretation of the cement/clinker diffraction profile, however, has several overriding problems [13]. For example, peak overlap is more of a rule rather than an exception. Additionally, the peaks themselves are somewhat variable in (2θ) position due to solid solution and substitution effects, that are difficult to unravel manually or by planimetric methods [14]. Today, most XRDs are equipped with automated (computerized) JCPDS search/match software that considerably eases the phase identification process. With careful calibration, XRD can also be used to quantify phases.

Sample preparation techniques

The above discussion focussing on the principles of different microstructural techniques that are available to the cement scientist sets the stage for examining the importance of sample preparation techniques, since each mode of analysis requires a specific preparation method.

XRD specimens are finely ground (around 5 μm) which allow random orientation. The technique uses the principle of characteristic planar reflections from a multitude of randomly oriented crystals [15]. The fineness of the ground clinker assumes some importance when comparing different samples, and can influence quantitative results.

In the case of "as-received" clinker/cement, the XRD pattern yields very little useful information due to the complexity of the material (peak overlap, peak shifts, etc.). Selective chemical dissolution of phases is necessary for accurate mineralogical identification.

Dissolution of interstitial phases can be obtained by the KOSH (KOH + sucrose) method on a portion of the sample [16]. This enables us to determine with reasonable accuracy the polymorphic form/s of C_3S and C_2S present in the sample. Another portion is treated with salicyclic acid + methanol to remove the silicate phases [16]; this simplifies the

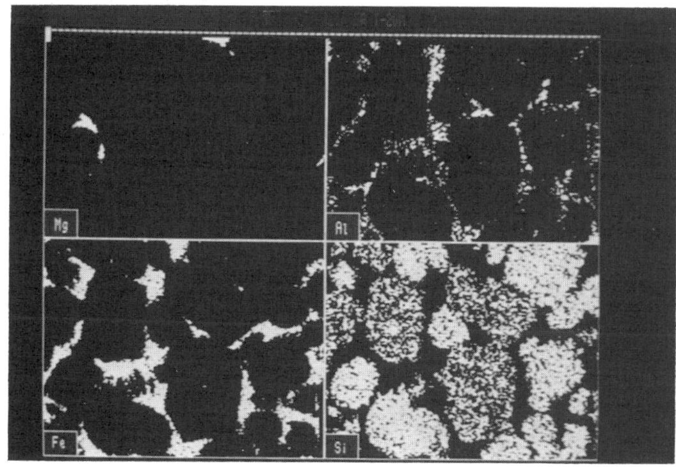

(a)

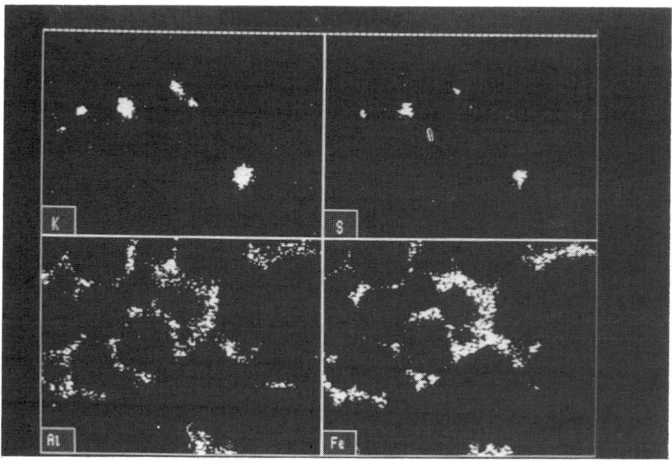

(b)

FIG. 6.--Elemental X-ray maps of a clinker showing Mg occurring as MgO in (a) and K and S as K_2SO_4 in (b)

identification of C3A polymorphs, alkali sulfates, periclase, and unconverted raw materials such as calcite and quartz, together with precise determination of the A/F ratio in the aluminoferrite phase. Figure 8 illustrates the advantage of selective chemical dissolution for identification of both major and minor phases in clinker. For cement, salicyclic acid + methanol treatment helps to enhance the effect of gypsum conversion to hemihydrate that commonly occurs in the cement grinding stage. Figure 9 illustrates that cement A has most of its calcium sulfate as hemihydrate (14.8°2θ), while B has much more retained as gypsum (11.7°2θ). In the first case the cement was ground in a hot mill or very low relative humidity, whereas the second one was ground in a cooler mill or much higher R.H. This information helps in predicting the rheological behaviour of cement, since the solubility of these two forms of calcium sulfate are quite different.

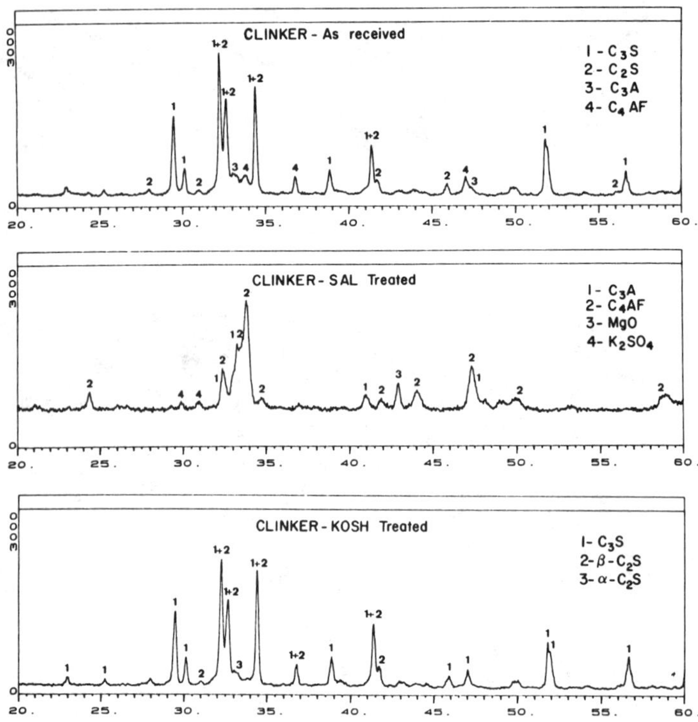

FIG. 8--XRD patterns of (a) as-received clinker (b) interstitial matrix phases removed and (c) silicate phases removed

Both fractured and flat polished specimens can be observed under the SEM, the only restriction is that cementitious substances have very low electrical and thermal conductivity. Therefore, it is necessary to apply colloidal graphite or silver paint on the side of the specimen to be examined. The specimen must also be coated with a thin layer of metallic (usually Au or Au-Pd) or carbon film to prevent surface charge accumulation under electron beam impingement. Compared to carbon evaporation on the surface, cold sputtering of metallic compounds is much faster, and minimizes thermal damage to the

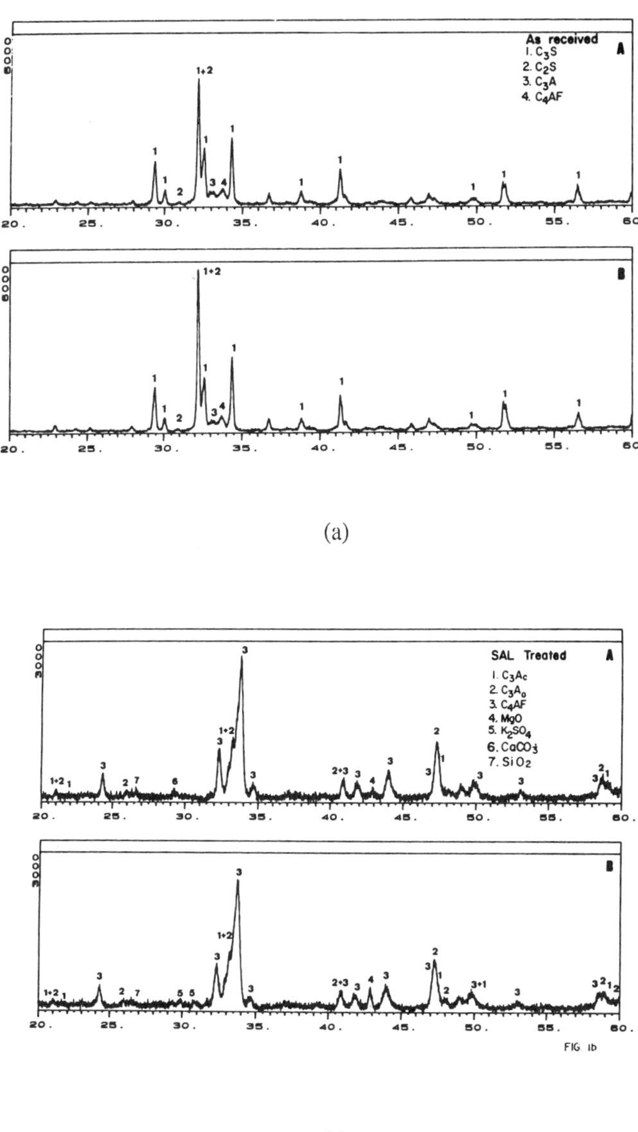

(b)

FIG. 7.--XRD traces of two (a) as-received and (b) salicyclic acid + methanol treated clinkers which yield nearly the same amount of C_3S, C_2S, C_3A and C_4AF from Bogue calculation. Their XRD patterns, however, show distinct differences

specimen. Specimen coating, however, can cause some absorption of incoming electrons and emitted soft x-rays from light elements, and possibly secondary fluorescence [17].

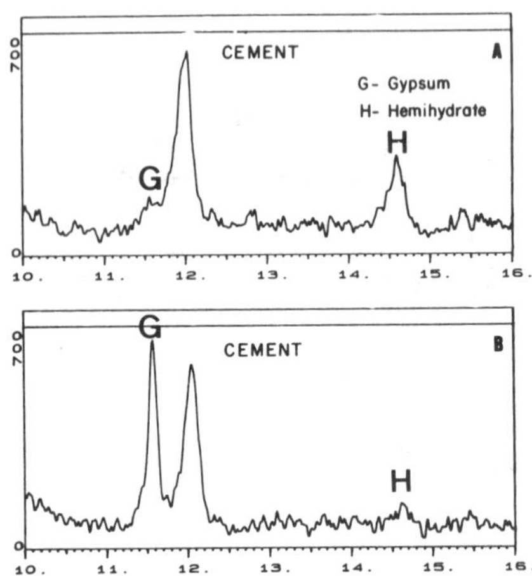

FIG. 9--XRD patterns of two salicyclic acid + methanol treated cements that were ground differently resulting in different proportions of gypsum and hemihydrate

The polished section preparation method for both optical microscopic examination and EDX microanalysis may be broken down into the following consecutive stages:

A few predried clinker nodules placed in a plastic mould, is resin impregnated. Different types of commercial resins are readily available for this purpose. A slice is then removed from the face of the impregnated section using a thin-bladed low-rpm circular saw to expose the clinker. This ensures minimum 'tearing' of grains from the surface, and white spirit, mineral or other non-aquous lubricant (in place of water) is used to prevent hydration. Rotary disc laps smeared with different grades of coarse-to-fine diamond abrasive pastes (ranging from 45 to 3 μm) are used sequentially to obtain a high quality 'even' surface polish. A final polish, not lasting more than a few minutes, is usually given with a1-0.5 μm diamond paste. Microscopic examination needs to be carried out at regular intervals during polishing so as not to overpolish the surface, which tends to 'pluck' out the weak phases [18].

For transmitted light microscopic examination of clinker, thin section, normally of 20-30 μm specimen thickness is required. This is made by mounting the clinker on a glass slide, followed by grinding it to the desired thickness. A thin cover slip is usually glued on to the exposed face to prevent hydration at a later stage.

That proper etching is extremely useful in bringing out the intricate details is demonstrated in Figure 10. These optical micrographs show two clinkers burned under different conditions which resulted in the development of two types of lamellar structure in C_2S. The sets of broad lamellas represents high firing temperature, followed by rapid cooling that gave rise to the formation of α-and β-C_2S without the intermediate α'-form. The closely knit cross-lamellar structure in the other clinker indicates the absence of the α-form as well as slower cooling. This example is provided basically to highlight the fact that proper sample preparation is a prerequisite for systematic optical microscopic study of clinkers, as it is in XRD, SEM, BEI and other systematic studies.

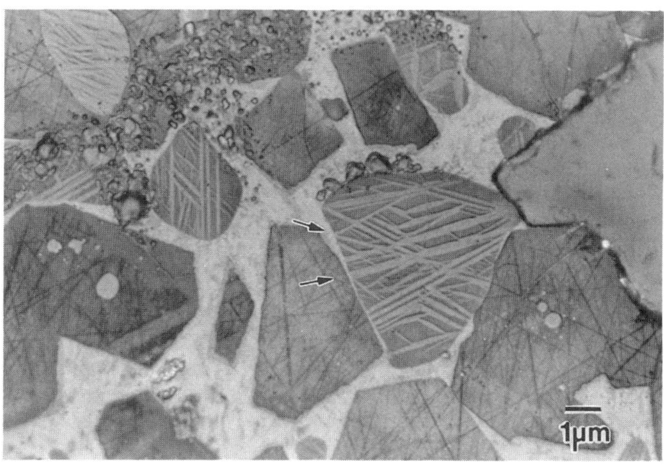

(a)

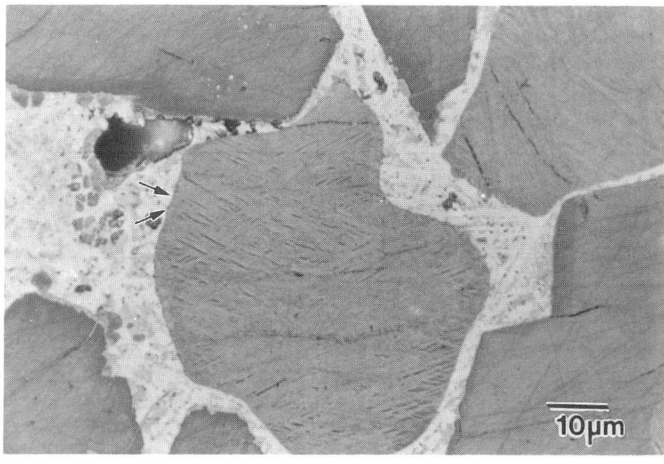

(b)

FIG. 10.--Optical micrographs of two clinkers succinic acid etched: (a) showing sets of broad lamellas, (b) cross-lamellar structure in C_2S

The present state-of-the art of energy dispersive x-ray microanalysis also rests predominantly on sample preparation. Sample effects, as such cannot be corrected for, and therefore, a good polished specimen surface is critically important for accurate quantitative analysis [19]. In SEM electron beam excitation, the small volume of specimen (whether fractured or flat polished) being analyzed must be homogeneous over the region in which the characteristic X-rays are produced. A grain must, therefore, have sufficient area and depth before it can be unambiguously analyzed. Additionally, the area should preferably be reasonably flat for several microns, and must be electrically surface conductive.

CASE STUDIES

The desired properties of cement partly depend on the physicochemical and mineralogical properties of clinker from which it is produced [20]. The grinding fineness and thermal history of the clinker are also very important. Additives added to the cement are known to change its properties significantly. With all the variables considered, the microstructural composition of the clinker is one of the several important parameters that must be within certain quality "windows". The following case studies illustrate the importance of clinker microstructure in evaluating the quality and predicting the performance of cement.

Case study 1

A Type 10 (ASTM Type I) cement produced in a coal fired dry-process kiln was periodically found to exhibit poor 28-day strength (Table 2), although the raw meal composition and firing conditions were kept constant, and their resultant physical and chemical characteristics were found to be very similar (Table 3). The Bogue compositions, however, were somewhat different (same table). This could be directly affected by the CaO value, or any other value for that matter.

TABLE 2--Compressive strength of two cements A and B from the same kiln

Compressive strength (MPa)	1 day	3 days	7 days	28 days	Strength level
Cement A	13.3	25.1	31.1	33.2	low
Cement B	12.9	23.3	27.0	39.5	normal

An in-depth comparative microstructural investigation of these two clinkers was undertaken in order to establish the cause of sporadic strength loss [21]. XRD analysis revealed the presence of $KC_2\bar{S}_3$ in the low-strength clinker (A). This was verified from SEM examination (Figure 11). From optical microscopy C_2S in this clinker was found to be exceptionally large (Figure 12). It was concluded that profusion of soluble alkali sulfate, and the delayed hydration of C_2S due to its large size are responsible for its low late-age strength. A host of workers [22] have reported that alkali sulfate in cement can have an adverse effect on the final strength of cement. Since large belite is a definite indication of long kiln residence time, its reduction was recommended, together with stringent control on kiln dust insufflation in order to lower the alkali sulfate content. This decreased the problem considerably [23].

TABLE 3--Physical and chemical analysis of cement A and B

Physical properties	Cement A	Cement B
Blaine fineness (m^2/kg)	336.6	336.0
Passing 45 μm (%)	88.0	86.4
Passing 75 μm (%)	98.4	98.5
Autoclave expansion (%)	0.001	0.001
Initial setting time (min)	135	145
Final setting time (min)	260	265
Air content (%)	9.4	9.6
False set-paste (%)	70.2	76.0
Chemical analysis		
LOI	2.5	2.9
SiO_2	20.9	20.7
Al_2O_3	4.7	4.7
Fe_2O_3	2.9	2.9
CaO	63.6	62.3
MgO	1.5	1.4
SO_3	3.1	3.2
Na_2O	0.4	0.4
K_2O	1.0	1.0
Na_2O equiv.	1.0	1.0
Free lime	0.3	0.3

Bogue composition

	C_3S	C_2S	C_3A	C_4AF
Cement A	55.52	18.04	7.55	8.82
Cement B	51.46	20.53	7.55	8.82

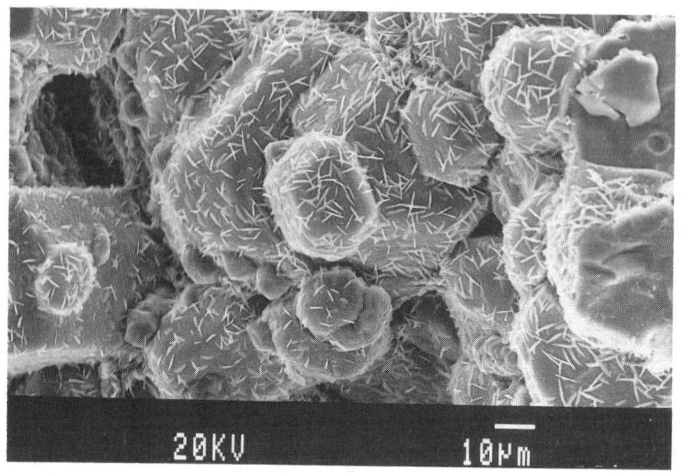

(a)

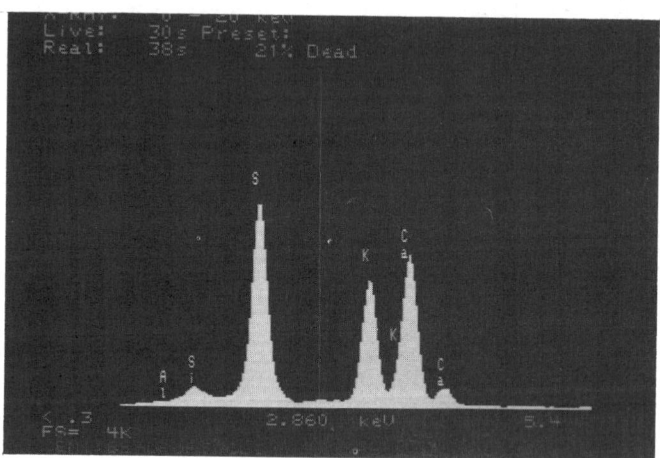

(b)

FIG. 11.--(a) Potassium-calcium sulfate crystals in a clinker. Its counterpart cement exhibits low 28-day strength. (b) EDX spectrum of $KC_2\overline{S}_3$

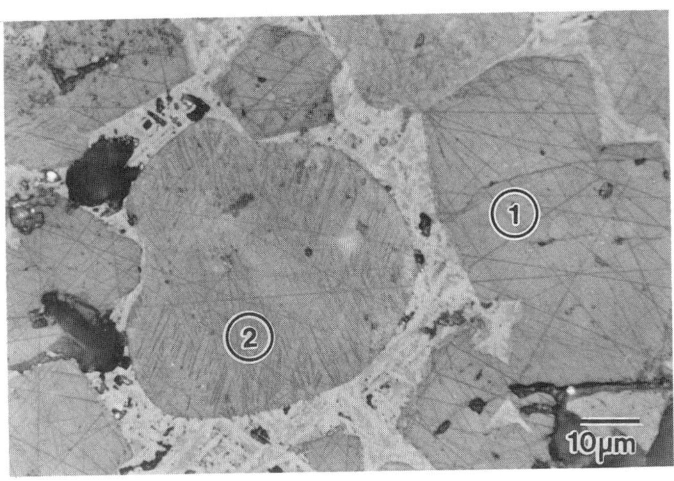

FIG. 12-- Large C_2S crystals in the same clinker. 1 = C_3S, 2 = C_2S

Case study 2

Industrialization can generate large quantities of wastes and by-products. An important aspect of achieving balanced industrial growth lies in the preservation of environmental quality.

An innovative approach to use limesludge, a waste from the PVC plant has been taken up by an Indian cement company. It is now using this waste limesludge as one of the principal raw materials in the manufacture of portland cement. The unusual nature of the raw meal, which also contains low grade limestone (Table 4) that is normally not suitable for cement manufacture, led to a microstructural investigation of its clinker quality [24].

TABLE 4--Raw mix and clinker phase composition

Raw	Limesludge	High grade limestone	Low grade limestone	Iron ore
mix	20-40%	2-14%	62-67%	1%
Clinker	C_3S	C_2S	C_3A	C_4AF
	61-65%	11-15%	8%	10-11%

Though the microstructure of the clinker did reveal certain inhomogeneities related to raw meal grinding, burning and cooling, the high compressive strength of this cement (Table 5) is attributable to its high C_3S content and small well-crystallized C_3S shown in Figure 13.

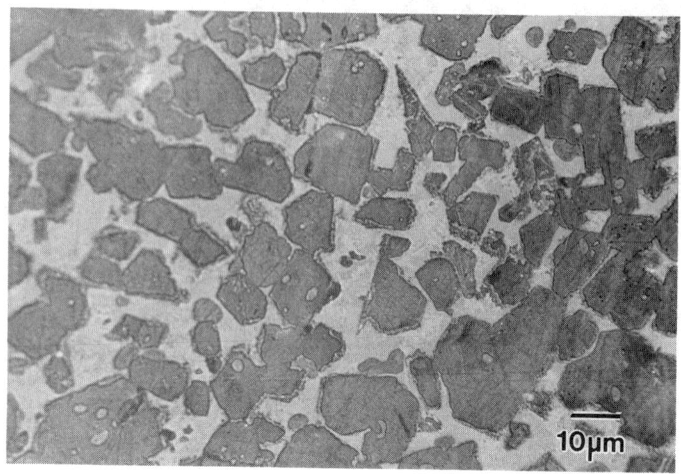

FIG. 13--Small well-crystallized C_3S in the clinker made from limesludge as one its raw meal components

TABLE 5--Properties of cement made from waste limesludge

Blaine fineness (m^2/kg)	270 - 280	
Initial setting time (min)	98	
final setting time (min)	145	
Compressive strength (MPa) (Indian Standards Institute Specification)	1 day	17.4
	3 days	28.1
	7 days	37.2
	28 days	48.8

Case study 3

With the objective of improving the burning process and reducing the clinker free lime content, modifications were carried out in a slurry grinding unit of a wet-process cement plant [25]. That the raw meal originally contained a high proportion of coarse siliceous grains was evident from the profusion of compact belite nests in the clinker when examined under the microscope (Figure 14). Based on this finding the ball charge in the mill was adjusted to improve the raw meal fineness, and thus improve raw meal burnability in the kiln. An instant improvement in the strength properties of cement was observed.

Case study 4

The influence of varying the SO_3 content on clinker microstructure and properties of corresponding cements was studied [26] using three commercial clinkers in which the SO_3 ranged from 0.62 to 1.28%. The alkali sulfate formed was apthitalite [$(NaK)_2$ SO_4]

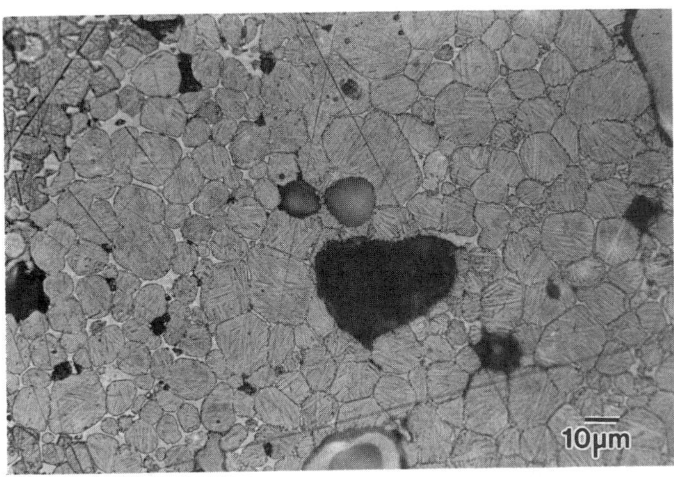

FIG. 14--Compact C_2S nest due to coarse siliceous particles in the raw meal

(see Figure 15). Coarsening of alite crystals and increase in grain idiomorphicity due to reduced melt viscosity was observed in the high-sulfate clinker (Figure 16). The secondary belite was a result of slow cooling. Differences in C_3A polymorphic forms in these clinkers were also noted from XRD analysis. The study concluded that cement strength development pattern and rheology are strongly dependent on the optimum SO_3 level in clinker. The cement manufacturer now makes every effort to maintain this optimum level and rapid cooling in order to produce better quality clinker.

Case study 5

The microstructure of clinkers from two types of kilns, one equipped with a two-stage suspension preheater and planetary cooler, and the other with a five-stage preheater, inclined calciner and grate cooler, both using the same source of raw materials and fuel, was examined [27] following conspicuous differences in strength and setting properties of cements made from clinkers burned in these kilns. It was demonstrated that the grinding method, kiln type and cooling procedure have a profound effect on the clinker microstructure, and therefore on the ultimate properties of cement. The closed circuit two-chamber mill with rotary turbo valve separator was found to perform more efficiently in terms of raw meal grinding compared to the open circuit ball mill with twin stationery air separators. The glassy interstitial matrix of the clinker cooled by grate cooler indicated rapid cooling, whereas the efficiency of the planetary cooler was lower due to difficulty in controlling its air supply. The microstructure of the clinker produced in the five-stage preheater kiln indicated better mix combinability and burnability.

Case study 6

In order to insure extended supply of high quality raw materials, i.e., to prolong the life of the quarry, cement producers are often forced to opt for alternate sources of lower quality/impure raw constituents, which if given the option, one would not normally use to manufacture cement. Such a situation arose in the case of a dry-process cement plant that

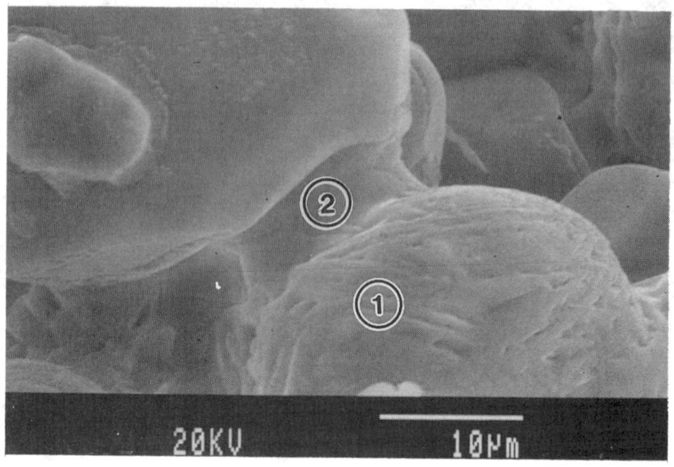

FIG. 15--Apthtalite $(NaK)_2$ SO_4 formation (2) between belite (1) grains in a high sulfate clinker

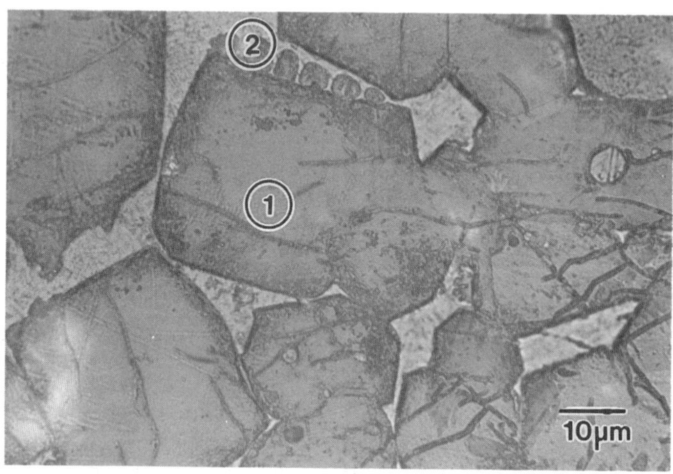

FIG. 16--Coarsening of C_3S crystals (1) and secondary C_2S (2) in this clinker

originally used silica as one of the raw meal components, but decided to replace it with sandstone. Considerable changes had to be made in the raw meal composition to produce the same cement type. A comparative clinker microstructure study was undertaken [28] to establish the differences that resulted from this change of raw meal.

TABLE 6--Physical properties and compressive strength of cements

Cement	% residue		Blaine fineness	Setting time (min)		Compressive strength (MPa)			
	45 μm	75 μm	(m^2/kg)	Initial	Final	1d	3d	7d	28d
With silica	89.5	99.0	330	182	299	9.9	20.7	32.4	44.3
With sandstone	89.9	99.0	326	171	286	10.1	19.3	29.4	38.9

Though the Blaine fineness of both these cements is nearly of the same order (Table 6), differences in setting time is evident, as is the compressive strength. Preponderance of alkali sulfate was noted in the sandstone-clinker. C_3S also appeared to have become coarser, and clinker microstructural features indicated a reducing condition in the kiln. Based on these findings, adjustments were later made in the burner pipe position, reported in another study [29], to reduce the generation of alkali sulfate, and thus improve the properties of cement.

CONCLUSIONS

Proper microscopic clinker examination provides invaluable textural and morphological information that are pertinent from the points of view of raw mix, clinkering parameters (pyroprocessing), and cooling conditions. The shape and size of C_3S and C_2S crystals, nature of grain boundary, C_2S lamellas, interstitial matrix phases, presence/absence of grain agglomeration - all reflect on the quality of the clinker. These, and other optical properties such the alite crystal long axis, short diameter of belite crystals, colour, pleochroism and birefringence give positive indication of the pyrotechnology.

Although a number of microstructural investigative techniques exist, optical microscopy still remains a highly useful, simple, and inexpensive tool for the cement technologist. Though it offers a relatively rapid means of diagnosing clinker, sample preparation for reflected light microscopy is elaborate, time consuming, and somewhat demanding. Nevertheless, it can reliably serve to obtain very useful information concerning the thermal history of the clinker. Although it does provide some definite answers, additional investigation using other methods, if available, is strongly recommended in order to obtain very precise results.

With its resolution limitations, and its inability to derive chemical composition, the optical microscope is gradually giving way to the SEM (with EDXA attachment) as its adjunct. Fractured and polished clinker specimens can be examined and analyzed simultaneously under the SEM. Coating the specimen to make it electrically conductive is a prerequisite to SEM examination. Improved methodology, selection of optimal experimental parameters, careful sample preparation and proper software manipulation enables the achievement of a high degree of accuracy in quantitative microprobe analysis of clinker phases by EDXA.

XRD is a very useful analytical tool for the cement scientist; its versatility lies in its ability to provide, in principle, direct means of obtaining mineralogical phase composition. Nonetheless, precise interpretation of cement and clinker diffractograms is an extremely arduous task for the diffractionist, attributable to peak shift due to solid solutions, peak overlap, and the possibility of the existence of more than one polymorph of a particular phase. Therefore selective dissolution of phases proves very useful for identification of major as well as minor compounds in cement and clinker.

REFERENCES

[1] Maxwell-Cook, P., Managing Editor, World Cement, Private communication, 1992.

[2] Barnes, P., Jeffery, J.W., and Sarkar, S.L., "Composition of Portland Cement Belites," Cement and Concrete Research, Vol. 8, 1978, pp. 559-564.

[3] Insley, H.H., "Structural Characteristics of Some Constituents of Portland Cement Clinker," Journal of Research of National Bureau of Standards, Vol. 25, 1940, pp. 295-307.

[4] Ellson, D.B., and Weymouth, J.H., "The Etching of Portland Cement Clinkers," CSIRO Chemical Research Laboratory Technical Paper, No. 5, Australia, 1968.

[5] Campbell, D.H., "Microscopical Examination and Interpretation of Portland Cement and Clinker," Construction Technology Laboratories, Skokie, 1986.

[6] Sarkar, S.L., "The Importance of Microstructure in Evaluating Concrete," Advances in Concrete Technology, V.M. Malhotra, Ed., Energy Mines and Resources, Canada, Ottawa, 1992, pp. 123-154.

[7] Skalny, J., Mander, J.E., and Meyerhoff, M.H., "SEM Study of Partially Dissolved Clinkers," Cement and Concrete Research, Vol. 5, 1975, pp. 119-128.

[8] Fundal, E., "The Description of a New Potassium - Alumina Phase in Ordinary Portland Cement," Proceedings of the 8th International Congress on Chemistry of Cement, Rio de Janeiro, Vol. II, 1986, pp. 139-145.

[9] Diamond, S., Young, J.F., and Lawrence, F.V., "Scanning Electron Microscopy - Energy Dispersive X-ray Analysis of Cement Constituents - Some Cautions," Cement and Concrete Research, Vol. 4, 1974, pp. 899-914.

[10] Schrivener, K.L., "The Microstructure of Concrete," Materials Science of Concrete, J.P. Skalny, Ed., American Ceramic Society, Westerville, 1989, pp. 127-162.

[11] Diamond, S., and Bonen, D., "A Chemical Image Analysis Method for Portland Cements," Materials Research Society Proceedings, Vol. 245, Boston, 1992, pp. 291-301.

[12] Sarkar, S.L., "X-ray Mapping - A Supplementary Tool in Clinker Phase Characterization," Cement and Concrete Research, Vol. 14, 1984, pp. 195-198.

[13] Sarkar, S.L., "Clinker Phase Identification by Automated X-ray Diffractometry - An Evaluation," Proceedings of the 10th International Conference on Cement Microscopy, San Antonio, 1988, pp. 30-41.

[14] Sarkar, S.L. and Roy, D.M., "QXRD in Cement and Clinker Phase Analysis: Its Progress and Limitations," Ibid, pp. 285-297.

[15] Klug, H.P. and Alexander, L.E., "X-ray Diffraction Procedures," John Wiley and Sons, New York, 1979.

[16 Gutteridge, W.A., "On the Dissolution of Interstitial Phases in Portland Cement", Cement and Concrete Research, Vol. 9, 1979, pp. 319-324.

[17] Chatterji, S., "Some Precautions That Must be Taken in the Scanning Electron Microscopic Study of Cements," Cement Technology, Vol. 2, 1971, pp. 5-9.

[18] Barnes, P., Moore, N.T. and Sarkar, S.L., "Sample Preparation Technique for Quantitative Energy Dispersive Analysis of Cementitious Materials in the Scanning Electron Microscope," Indian Concrete Journal, August 1977, pp. 251-252.

[19] Sarkar, S.L., "EDXA in the Analysis of Cement," Advances in Cement Technology, S.N. Ghosh, Ed., Pergamon Press, Oxford, 1983, pp. 711-732.

[20] Tagnit-Hamou, A. and Sarkar, S.L., "Applications of SEM/EDXA in Cement Clinker Investigation," Proceedings of the XIIth International Congress for Electron Microscopy, Seattle, 1990, pp. 870-871.

[21] Sarkar, S.L., "Microstructural Investigation of Strength Loss in a Type 10 Cement," Proceedings of the 11th International Conference on Cement Microscopy, New Orleans, 1989, pp. 101-114.

[22] Jawed, I. and Skalny, J., "Alkalies in Cement: A Review. I. Forms of Alkalies and Their Effect on Clinker Formation," Cement and Concrete Research, Vol. 7, 1977, pp.719-730.

[23] Samet, B., Tagnit-Hamou, A., and Sarkar, S.L., "The Effect of Minor Kiln Variations Upon Clinker Quality," World Cement, June 1992, pp. 12-17.

[24] Sarkar, S.L., Brahma, N.C., and Dua, B.S., "Utilization of Industrial Wastes in the Manufacture of Cement - A Case Study," Proceedings of the 12th International Conference on Cement Microscopy, Vancouver, 1990, pp. 54-66.

[25] Sarkar, S.L., Seshadri, T.R. and Rao, D.B.N., "The Adverse Effects of Particle Size Distribution of Kiln Feed on Pyroprocessing," World Cement, June 1990, pp. 245-248.

[26] Tagnit-Hamou, A. and Sarkar, S.L., "The Influence of Varying Sulphur Content on the Microstructure of Commercial Clinkers and the Properties of Cement," World Cement, September 1990, pp. 389-393.

[27] Tagnit-Hamou, A. and Sarkar, S.L., "Microstructural Study of Clinkers from Two Types of Kilns Using the Same Raw Materials," Proceedings of the 13th International Conference on Cement Microscopy, Tampa, 1991, pp. 160-176.

[28] Sarkar, S.L. and Tagnit-Hamou, A., "Replacement of Silica by Sandstone in a Raw Meal - Its Effect on the Clinker Microstructure," Ibid, pp. 1-16.

[29] Samet, B., Bonen, D., and Sarkar, S.L., "Effect of Burning Conditions on the Clinker Microstructure and Paste Characteristics," Proceedings of the 15th International Conference on Cement Microscopy, Dallas, 1993, pp. 73-88.

Willis A. Weigand[1]

PROGRESS TOWARD A STANDARD PROCEDURE FOR POINT-COUNTING THE PHASES IN CEMENT CLINKER WITH REFLECTED LIGHT MICROSCOPY

REFERENCE: Weigand, W. A., **"Progress Toward a Standard Procedure for Point-Counting the Phases in Cement Clinker with Reflected Light Microscopy,"** Petrography of Cementitious Materials, ASTM STP 1215, Sharon M. DeHayes and David Stark, Eds., American Society for Testing and Materials, Philadelphia, 1994.

ABSTRACT: Light microscopy can be used to gain insight into the cement manufacturing process and to guide the manufacturer in the production process. By quantitative phase determination, the microscopist can determine how much of each phase is present as well as its morphology.

The ASTM task group C01.23.02 on Microscopical Methods, under the C01.23 subcommittee on Compositional Analysis, has been developing a Standard Practice for point counting of cement microstructure using reflected light microscopy. The task group is continuing with round robin testing. The inter-laboratory study results for one SRM indicate that counting 3 000 points is a good compromise between effort and accuracy. Intra-laboratory precision showed a need for a uniform level of training; therefore, a video tape has been made for training microscopists.

KEYWORDS: Reflected light microscopy, cement, clinker, quantitative phase analysis

---

INTRODUCTION

The production of portland cement is a complex chemical reaction of materials such as limestone, sand, and clay at high temperatures. The composition of the finished product is thought to control the behavior of the cement which includes set time, strength, and durability. The cement plant chemist has a number of analytical tools available to determine the quantity and phases present in the cement clinker. The determination of the quantity and phases in the cement clinker may allow

---

[1]11100 Crossland Dr., Austin, TX 78726-1324

the plant chemist to predict the performance of the portland cement.

There are four common methods of determining phase abundance. Quantitative X-ray diffraction (QXRD), scanning electron microscopy (SEM), Bogue potential calculations from chemical analysis, and optical microscopy, especially reflected light microscopy, are some of the analytical methods available. Reflected light microscopy and QXRD are methods which directly measure the amount of each phase present in the cement clinker. QXRD methods are being developed in another task group under subcommittee C01.23 and results of that task group have been reported in an earlier paper [1]. The goal of this task group is to develop a Standard Practice for point counting of cement microstructure. This paper will present: (1) a brief history of the task group, (2) a summary of the Standard Practice, and (3) results of one of the round robin tests that was conducted by the group.

HISTORY OF THE TASK GROUP

Optical microscopy of cement was described as early as 1887 by Le Chatelier [2]. Over the last one hundred years, much research has been conducted on cement using reflected light microscopy. As the use of the microscope spread, it became clear that a standard method for reflected light microscopy would be helpful to have a clearly defined way for identifying and counting the phases of clinker.

A task group was formed in the early 1980's under subcommittee C01.23 to develop a method for cement microscopists to follow when performing point counts. In one of the first meetings, results of a round robin were discussed. In that round robin the point counts ranged from 2 000 to 12 100. In 1987, another group was assembled, chaired by Dr. Don Campbell, to incorporate the task group's general consensus in a method for subsequent round robin tests.

DEVELOPMENT OF THE STANDARD PRACTICE

Typically, the cement microscopist encounters two decision points when performing a point count on cement clinker. These are the correct identification of the phase being observed and how to count that phase. The identification of a particular phase is determined by the subjective judgment of the microscopist and the experience with identification of clinker phases. Normally, when a microscopist is conducting a point count, the polished clinker is etched with some type of etchant to ease in the identification and counting of the various phases. These etchants enhance either the alite (C_3S), belite (C_2S), or aluminate (C_3A) phases[2]. The silicate phases can be etched with dilute salicylic acid, ammonium chloride, isopropyl alcohol/water solution, and probably the most commonly used etchant, Nital, which is a mixture of nitric acid in a low molecular weight alcohol such as ethyl or isopropyl.

[2]Cement chemist's notation: C=CaO, S=SiO_2, A=Al_2O_3, F=Fe_2O_3

The aluminate phase can be etched with water or solutions of sodium or potassium hydroxide [2].

Once the phases have been identified, the second decision point is the correct counting method to determine their abundance. Generally, the counting method uses the microscope cross-hair reticules to count the phases as they fall under the cross-hairs as a clinker surface is scanned or by counting the phases in the quadrants formed by the cross-hairs. Either method will produce satisfactory results as long as a consistent method is utilized.

The importance of determining the minimum number of points to produce a valid result is critical because physically counting four major types of phases in a selected cement clinker is time consuming and sometimes visually difficult. As was noted earlier, in some of the first round robins conducted by the task group, point counts ranged from 2 000 to 12 000. Much of the work by this task group has been on these two issues of identification and number of point counts required for a statistically valid determination. The remainder of this paper describes the standard practice developed in the task group and the results of a round robin which was conducted to determine the minimum number of point counts to be required and the inter- and intra-laboratory variability in a selected cement clinker.

Three typical production clinkers, identified as SRM 2686, SRM 2687, and SRM 2688, are available from the National Institute of Standards and Technology (NIST). These standard reference materials(SRM) were characterized by QXRD and reflected light microscopy at the Construction Technology Laboratories and by SEM with image analysis at NIST [3]. These SRM's have been used in the round robin testing conducted by the microscopy task group. The round robin results presented in this paper are from SRM 2686 [4].

Summary of Method

The method under development describes the major and minor clinker phases which the microscopist will be identifying during the point count. The list is not extensive nor is it expected that the microscopist will need to count all phases identified. This method describes the apparatus needed to conduct a point count indicating which equipment is absolutely necessary and which equipment is optional. Types of etchants to use are suggested, but the choice is left up to the microscopist. A section on sample preparation is included to guide the microscopist in selecting and mounting the clinker in epoxy and in producing the best surface for point counts. Each of these steps is carefully detailed to reduce sources of error in sample preparation.

In the counting procedure, reticles with multiple grid points are recommended and a magnification chosen such that adjacent reticle grid points do not fall on the same crystal. This will generally require magnifications from 200 to 500x. The sample is placed upon either a mechanical or electrically driven microscope stage. A starting point on the sample is then chosen. The clinker phases which fall under the cross-hairs of the reticle are counted and all phases which occur in that particular field of view are tallied. The stage is then moved to another view and those phases in the new field of view are tallied. This procedure continues until 3 000 to 4 000 points are recorded. The volume fraction of each phase in the sample is calculated by dividing the number of points counted for each phase by the total number of points counted and multiplying this number by 100 to yield the phase content in volume per cent. This was the method followed in the round robin results shown.

Round Robin Testing

All of the SRM's have been counted at some time during the course of the last five years of the task group. However, the most extensive round robin was conducted on SRM 8486 in which the task group members were asked to conduct point count analyses using 3 000, 4 000, and 5 000 data points per sample. Using this same sample the members were also asked to determine their intra-lab results, again utilizing 3 000, 4 000, and 5 000 point counts. The draft method from the standard practice was to be used, although the method was not in its final form at that time. The X-ray fluorescence analysis is shown in Table 1 and the Bogue potential analysis is shown in Table 2.

TABLE 1--Chemical analysis by x-ray fluorescence (wt %) [4]

Analyte	wt %
SiO_2	22.5
Al_2O_3	4.7
Fe_2O_3	3.6
CaO	63.4
MgO	4.7
SO_3	0.27
Na_2O	0.1
K_2O	0.42
TiO_2	0.25
P_2O_5	0.06
Mn_2O_3	0.1
SrO	0.05
Loss on Ignition	0.16
Total	100.28

Table 2--Bogue potential compounds (%)

Compound	%
C_3S	48
C_2S	28
C_3A	7
C_4AF	11

Seven laboratories participated in the inter-laboratory study and three of the seven labs participated in the intra-laboratory study. Table 3 shows the results of the inter-laboratory study at 3 000, 4 000, and 5 000 point counts. The average, standard deviation, and coefficient of variation are shown for the four major phases obtained by the seven laboratories.

Table 3--Results of the inter-laboratory study of SRM-8486

	C_3S	C_2S	C_3A	C_4AF
3 000 Point Count				
Average	52.9	22.5	2.9	10.3
SD	5.7	3.9	1.3	1.4
CofV	10.7	17.3	44.8	13.3
4 000 Point Count				
Average	53.8	22.8	3.2	10.3
SD	4.8	4.2	1.5	1.5
CofV	9.0	18.4	47.3	14.7
5 000 Point Count				
Average	53.3	22.8	3.0	10.5
SD	5.4	3.7	1.2	2.0
CofV	10.1	16.1	41.5	19

Table 4 shows the results of the the intra-laboratory study in which three laboratories returned data. Each of these labs conducted 3 000, 4 000, and 5 000 point count analyses.

Table 4--Results of intra-laboratory study of SRM-8486

3 000 Counts	C_3S	C_2S	C_3A	C_4AF	4 000 Counts	C_3S	C_2S	C_3A	C_4AF
Lab 3					Lab 3				
Average	50.4	26.8	2.8	11.7	Average	49.2	28.1	2.8	11.5
SD	2.6	5.5	2.1	0.6	SD	2.9	6.1	2.0	0.4
CofV	5.1	20.7	75	4.9	CofV	5.9	21.8	70.9	3.6
Lab 4					Lab 4				
Average	58.9	18.7	1.9	12.6	Average	58.3	19.2	1.9	12.7
SD	3.4	2.5	0.4	1.7	SD	2.3	2.7	0.4	1.0
CofV	5.8	13.4	20	13.7	CofV	4.0	14	18.7	7.9
Lab 6					Lab 6				
Average	54.4	20	4.5	9.8	Average	54.4	19.9	4.5	10
SD	0.5	0.2	0.3	0.6	SD	0.4	0.2	0.1	0.7
CofV	1.0	1.1	5.8	6.5	CofV	0.6	0.8	2.7	6.7
5 000 Counts									
Lab 3									
Average	49.7	27.5	2.6	11.6					
SD	3.2	6.3	2.0	1.0					
CofV	6.5	22.8	77	9.0					
Lab 4									
Average	58.2	19.1	1.9	12.8					
SD	2.0	2.3	0.4	0.5					
CofV	3.4	11.9	19.2	4.0					
Lab 6									
Average	54.4	19.9	4.5	9.9					
SD	0.4	0.1	0.1	0.6					
CofV	0.7	0.7	2.2	6.0					

RESULTS AND DISCUSSION

The results of the inter-laboratory study on SRM-8486 suggest that about 3 000 points provides sufficient data to arrive at results which represent a minimization of time by the microscopist and a relatively accurate determination of the major phases in the clinker. Hofmanner [5] has shown that the absolute error at the 96% confidence level is:

$$\delta = 2.0235 \sqrt{\frac{P(100-P)}{N}} \quad (1)$$

Where (delta) is the absolute measuring error in percentage for a given constituent, P is the phase percentage, and N is the total number of points counted. Any error percentage deemed acceptable can be used to calculate the required number of points, N, by the following formula:

$$N = \left(\frac{2.0235}{\delta}\right)^2 \times (100 - P)\,P \qquad (2)$$

If P is chosen to be 50%, and subsequently substituted into formula 2, a value of 3 000 is obtained at a delta of about 1.8 percent.

The averages of the abundance of each major phase in the clinker are very closely grouped regardless of the number of points counted. The standard deviation of the C_3S determination averaged over the 3 000, 4 000, and 5 000 points is 0.4. For the C_2S the standard deviation is 0.2, and for C_3A and C_4AF, 0.2 and 0.1, respectively. These data further suggest that point counts beyond about 3 000 do not significantly change results.

Compared to the Bogue calculation, the point count results produce a C_3S value about 5 percent higher, the C_2S about 5 percent lower, the C_3A 4 percent lower, and about one-half percent lower with the C_4AF. Figure 1 shows this graphically. The 4 percentage points in the C_3A determination represents more than one-half of the value calculated from the Bogue equation. This result may be due to the inability of the microscopist to determine all of the C_3A present in the clinker. The ability to discern C_3A is highly dependent upon the morphology of the aluminate phase, for example, whether it is in a larger tablet form or highly micro-crystalline. It is much easier for the microscopist to see and count the larger tablet form than the very small crystalline form. This result compares to the results obtained by the QXRD task group in which the results of the optical microscopy were about one-half of the value reported by QXRD [1].

The intra-laboratory results show the variability within each laboratory. In this case, only three labs reported results. Based upon the coefficient of variation it appears that there are some differences within laboratories. These differences may be due to the experience level of the microscopist, preparation of the sample, standardization of the counting technique, or misidentification of the phases. The task group did not specify the experience level or training requirements of the microscopist performing this round robin, so these results represent a broad range of experience.

The intra-laboratory study also shows that increasing the number of point counts does not significantly change results as the standard deviations do not significantly improve. This was also the case in the inter-laboratory study. Based upon these results the task group has determined that approximately 3 000 points provide sufficient data for an analysis. This minimizes time required per analysis and provides the microscopist a measure of confidence in the analysis.

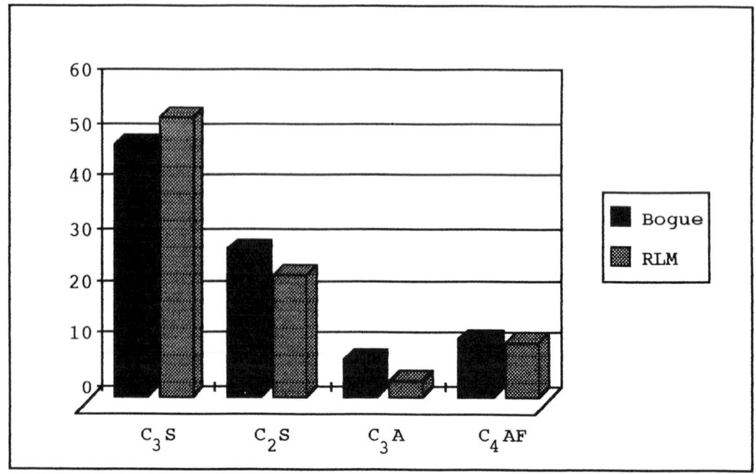

FIG. 1--Comparison of Bogue and reflected light microscopy(RLM) results of SRM-8486 (3 000 points)

Based upon the intra-laboratory results, the task group has also determined that training will be an important aspect in conducting a point count analysis. To that end, the task group has made available a video tape with narration of several different cement clinkers in which the microscopist can study the clinker phases.

The task group is currently conducting a round robin on SRM-8488. The results of this round robin will be used to determine whether the procedure as currently written provides enough needed information for the microscopist to conduct a point count of portland cement clinker. Future work by the task group will be to determine precision and bias of the standard practice.

CONCLUSIONS

The determination of phases in a portland cement clinker by optical microscopy requires a consistent method to produce consistent results. The publication of a standard practice by ASTM is a step toward providing the required information to those interested in performing this type of analysis. Results of a round robin are compared to Bogue results obtained from X-ray fluorescence data. The ASTM task group on microscopy will continue to be active in this area and is available for assistance for those interested in this technique.

Acknowledgments

This paper is a summary of a small part of the work of Task Group C01.23.02. The results shown could not have been done without the hard work of the members of this task group and the previous chairmen.

REFERENCES

[1] Struble, L. J., "Quantitative Phase Analysis of Clinker Using X-Ray Diffraction," *Cement, Concrete and Aggregates*, Vol. 13, No. 2, Winter 1991, pp. 97-102.

[2] Campbell, D. H., *Microscopical Examination and Interpretation of Portland Cement and Clinker*, Construction Technology Laboratories, Skokie, IL, 1986.

[3] Kanare, H. M., "Production of Portland Cement Clinker Phase Abundance Standard Reference Materials," final report, CTL Project No. CRA012-840, Construction Technology Laboratories, Skokie, IL, 1987.

[4] Rasberry, S. D., "Report of Investigation, Reference Materials 8486, 8487, 8488, Portland Cement Clinker," National Institute of Standards and Technology, May, 1989.

[5] Hofmanner, F., Microstructure of Portland Cement Clinker, Holderbank Management & Consulting Ltd., Switzerland, 1973, 48 pp.

Dale P. Bentz[1] and Paul E. Stutzman[1]

SEM ANALYSIS AND COMPUTER MODELLING OF HYDRATION OF PORTLAND CEMENT PARTICLES

REFERENCE: Bentz, D. P. and Stutzman, P. E., **"SEM Analysis and Computer Modelling of Hydration of Portland Cement Particles,"** Petrography of Cementitious Materials, ASTM STP 1215, Sharon M. DeHayes and David Stark, Eds., American Society for Testing and Materials, Philadelphia, 1994.

ABSTRACT: Characterization of cement particles is complicated due to their wide size range, complex shapes, and multi-phase nature. Accurate characterization should allow for better prediction of cement performance and more realistic modelling of cement microstructural development. This paper presents a technique, based on scanning electron microscopy and digital image processing, for obtaining two-dimensional digital images of actual portland cement particles in which all major phases are identified. By combining backscattered electron and X-ray images, an image segmented into the major cement phases may be created. These images can be analyzed to determine any number of quantitative measures such as phase fractions or phase perimeters. The technique has also been successfully utilized in obtaining realistic starting images for input into a two-dimensional cement microstructure model which simulates the hydration process.

KEYWORDS: building technology, cement particles, characterization, hydration, image processing, interfacial zone, microstructure, phase analysis, scanning electron microscopy, simulation, X-ray images.

The processes by which cement paste transforms from a viscous suspension into a rigid solid must be understood if the performance of concrete and other cement-based materials is to be reliably predicted from the properties of their constituents. While the hydration reactions and mechanisms have not been completely elucidated, the topic is further complicated by a lack of a quantitative description of the anhydrous cement particles. Both the bulk composition of a cement and the spatial distribution of the various phases will influence the temporal properties of a hydrating system. Quantification of the initial cement particles is thus seen as an important step in developing scientifically-based relationships between cement microstructure and resultant properties such as strength and durability. Additionally, such quantification could serve as a valuable quality control technique for the cement production industry, ultimately leading to a new

---

[1] Chemical engineer and physical scientist respectively, Building Materials Division, Building and Fire Research Laboratory, National Institute of Standards and Technology, U.S. Department of Commerce, Gaithersburg, MD 20899.

generation of cement and concrete standards.

In recent years, the application of scanning electron microscopy (SEM) to characterizing cement clinker and ground cements has increased. SEM and X-ray microanalysis have been utilized to identify the four major phases in portland cement clinker [1]. Scrivener has applied similar techniques to determine the distribution of silicates and interstitial phases in cement grains [2] and compared the results to the rates of heat release during hydration. Bonen and Diamond have analyzed the same cement clinker ground in both a ball mill and a high pressure roller mill, using SEM and X-ray analysis to classify the predominant phase found in each cement particle [3]. Additionally, the particles were characterized on the basis of aspect ratio and shape factor using image analysis techniques.

Another recent development in the field of cement research is the use of computer modelling to relate microstructure to properties. Often, these models are digital-image-based, with a two or three-dimensional digital image being the basis for the microstructure model [4]. The underlying image structure allows for the efficient computation of properties such as ionic diffusivity or elastic modulus [5]. While simplifications such as using circular particles can be made, these models can accept actual images of cement particles as input. It was therefore of interest to develop techniques to obtain images of cement particles in which each pixel (element) in the image has been classified as a phase of portland cement. This paper details the technique developed to obtain two-dimensional images of the real anhydrous cement particles and provides examples of using these images as input into a computer model of the microstructural development of hydrating cement paste.

EXPERIMENTAL METHOD

Sample Preparation

Proper sample preparation is critical to the successful imaging of fine cement particles. Difficulties encountered in preparing polished sections of cement include the elimination of scratches, edge rounding, surface relief, and grain plucking.

To prepare a specimen for viewing in the SEM, about 25 grams of the cement powder of interest are blended with an epoxy resin[2] to form an almost dry paste. The epoxy resin has a viscosity close to that of water so that this specimen will be similar to a suspension of cement particles in water. The paste is pressed into a sample mold and cured at 60 °C for 24 hours. The cured specimen is cut using a low-speed diamond wafering saw, cutting first a layer from the outer surface and then about 10 mm into the sample. This two-cut procedure produces parallel faces, minimizing the need to refocus as one traverses a specimen.

Saw marks are easily removed by dry grinding with 400 grit followed by 600 grit sandpaper. Final polishing is done on a lap wheel with 6, 3, 1, and 0.25 μm diamond paste for about 30 seconds each. We have been able to produce low relief, fine polishes using a lint-free

---

[2] EpoTek 301, Epoxy Technology Corporation, Billerica, Mass. This and other tradenames and company products are identified to adequately specify the experimental procedure. In no case does such identification imply recommendation or endorsement by the National Institute of Standards and Technology, nor does it imply that the products are necessarily the best available for the purpose.

polyester SEM cleaning cloth[3] and diamond paste on top of Texmet paper. The specimen is cleaned after each polishing stage by gently wiping on a clean cloth; residual polishing compound is removed with ethanol after the final polishing stage. Repeated cleaning with ethanol is avoided because it may soften the epoxy and increase the chance of grain plucking. The specimen is then coated with carbon to provide a conductive surface for viewing in the SEM.

SEM Imaging

When viewed in the SEM, signals emitted from the sample as a result of the specimen-primary electron beam interaction include backscattered electrons (BE) and X-rays. For this study, an accelerating voltage of 12 kV and probe current of about 2 nA was used for collecting the BE images, while probe currents of about 10 nA were used for X-ray imaging.

In the BE images, brightness is proportional to the average atomic number ($\bar{Z}$) of a phase. For the major phases present in portland cement, the phases from brightest to darkest are tetracalcium aluminoferrite (C_4AF), tricalcium silicate (C_3S), tricalcium aluminate (C_3A) and dicalcium silicate (C_2S), gypsum, and the resin-filled voids. Because the BE signal is weak, BE images are inherently noisy so that often the noise is reduced by averaging images of the same field of view. Phase identification can then be attempted by segmenting the BE image based on analysis of the greylevel histogram (a histogram of the number of pixels in the image assigned to each greylevel or intensity). While some phase segmentation is possible, a completely accurate separation cannot be based solely on the BE image since several of the phases (such as C_3A and C_2S) have similar intensities ($\bar{Z}$ values) even though they differ widely in chemical composition.

To supplement the information content of the BE image, X-ray images are obtained for the elements calcium, iron, aluminum, and sulfur. A typical X-ray spectrum from an energy dispersive X-ray detector appears in Figure 1 as a plot of signal intensity vs energy level. X-ray images are created by slowly scanning the specimen while measuring X-ray counts within energy windows encompassing a peak for the element of interest. Typically, about three hours of scan time is required to obtain a set of 512*400 X-ray images. These images of X-ray signal intensity can be processed and analyzed in the same manner as the BE image, and can even be combined with the BE image.

Image Processing

Figure 2 shows the BE image and the four X-ray images for an ASTM Type I portland cement. In these images, each pixel is about 0.5 μm on a side, which is near the resolution limit for X-ray images. For the iron, aluminum, and sulfur X-ray images, the threshold levels used to produce binary images were determined by finding a local minimum in the greylevel histogram for each image. For calcium, the X-ray image is segmented into four levels, indicating high, medium, low, and no calcium content, depending on the local (pixel) intensity of the X-ray signal. By combining these four X-ray images with the BE image, each of the four main clinker phases of a portland cement along with gypsum may be distinguished. For instance, the presence of iron in the X-ray image indicates the tetracalcium aluminoferrite phase while the presence of aluminum but not iron indicates the tricalcium aluminate phase. Similarly, gypsum is indicated by the presence of sulfur. Here, we should note that all sulfates, including the sodium and potassium sulfates, are assigned to be gypsum, since the microstructure model

---

[3] Lint-free cleaning cloths are available from most SEM suppliers.

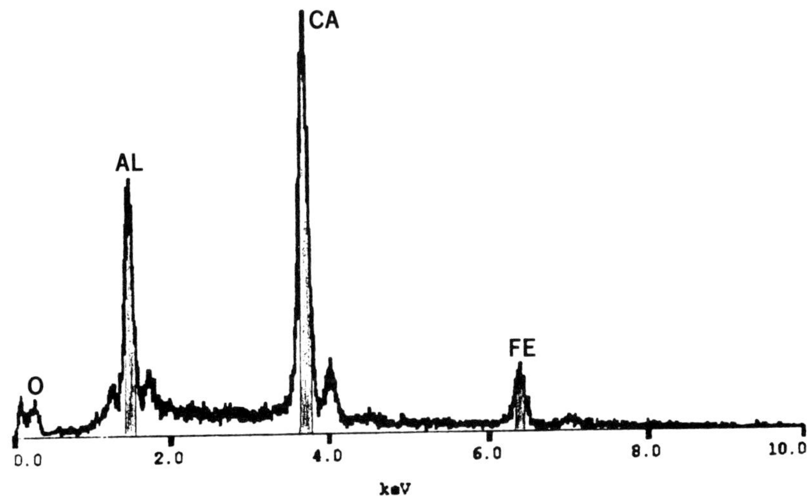

Figure 1. X-ray spectrum of tetracalcium aluminoferrite indicating a composition of calcium (Ca), iron (Fe), aluminum (Al), and oxygen (O). The vertical bars indicate the peak window region used for generating an X-ray image.

described later assumes all sulfates will react with aluminates to form ettringite and monosulfoaluminate. If further segmentation were required, however, the X-ray images for sodium and potassium could be used to distinguish the various alkali sulfates from gypsum. Finally, the tricalcium and dicalcium silicate are identified based on the intensity of the BE and calcium X-ray images, where the brighter areas in these images correspond to the tricalcium silicate phases. If the calcium X-ray and BE images are insufficient to separate the silicate phases, an X-ray image for silicon is collected to process along with these two images. Here, segmentation can be based on the calcium/silicon ratio, which should be higher for C_3S than for C_2S.

The segmented image produced in this fashion still contains noise and some pixels which are not porosity but which have not been assigned to one of the solid phases. To sharpen the phase distinction and eliminate the noise, the image is filtered using a median filter. Here, each "solid" pixel in the image is reassigned to be the phase occupied by the majority of its neighbors, excluding (resin-filled) porosity, if this majority exceeds a preset limit. Applying these processing steps to the images in Figure 2 results in the final image shown in Figure 3. The complexity of both shape and phase distribution for the cement particles shown in Figure 3 suggests that the artificial computer generation of such particles would be a formidable task, reinforcing the importance of the experimental technique described herein.

Images like those shown in Figure 3 can be further processed in a number of ways. The shapes and phase distributions of the individual cement particles can be stored in a database using a line segment encoding technique [6] and used to create computer-assembled images representing a given cement at a variety of water-to-cement (w/c) ratios, where the porosity is now assumed to be filled with water. The areas of the individual cement particles also can be assessed, although converting this two-dimensional size distribution to a true three-dimensional particle size distribution is possible only when assumptions

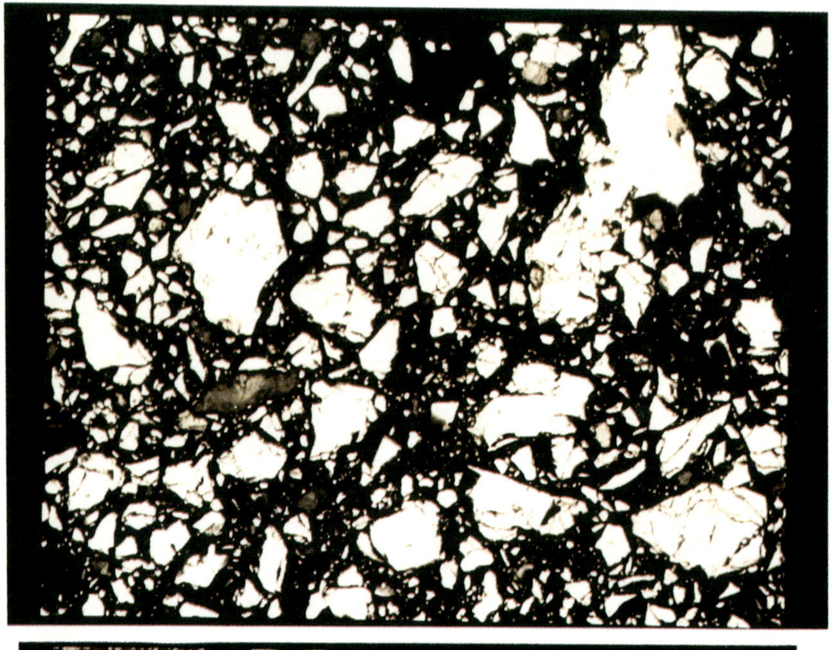

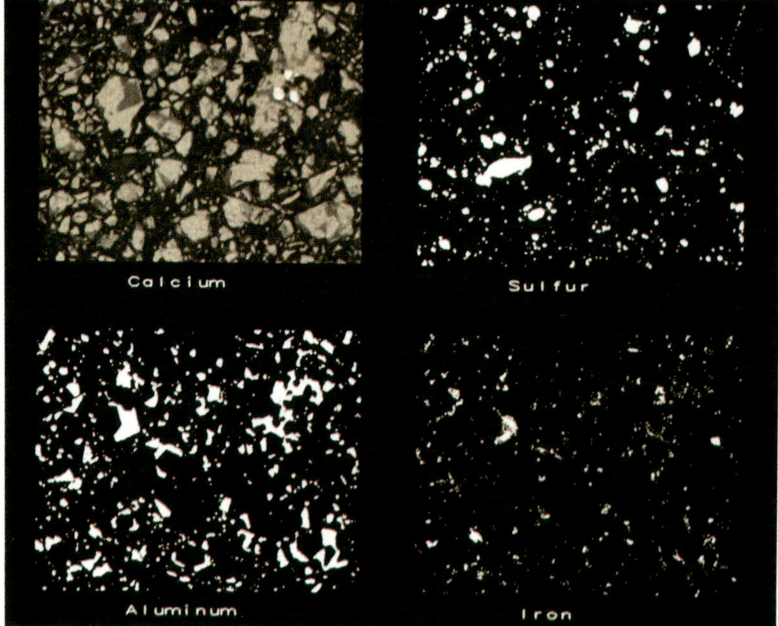

Figure 2. Backscattered electron (top) and X-ray images (bottom) for a real Type I portland cement.

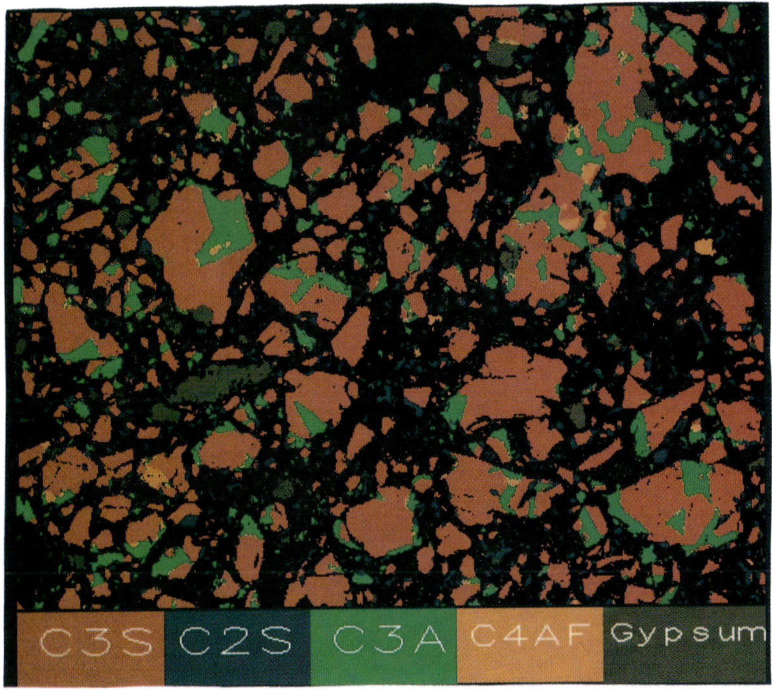

Figure 3. Final 256*199 μm (512*398 pixel) image of real cement particle microstructure (w/c=0.36) for images shown in Figure 2. Phase colors are as indicated in color bars at the bottom of the image.

concerning particle geometry (such as assuming spherical particles) are made [7]. The bulk area fractions, which should correspond to volume fractions [7], can be computed by simply counting the number of pixels assigned to each phase. These fractions can then be compared to those computed on a volume basis by applying the Bogue calculation to the oxide composition of the cement. Since cement hydration is critically dependent on the contact surface between water and the cement particles, the perimeter or surface fraction of each of the phases is also a quantity of interest. This measure can be determined by counting those pixel edges separating porosity and solid pixels for each solid phase.

CEMENT HYDRATION MICROSTRUCTURE MODEL

At the National Institute of Standards and Technology (NIST), images such as the one in Figure 3 have been used extensively as input to a digital-image-based cement hydration microstructure model. The goal of the model is to simulate the microstructure of cement paste as it hydrates. Hydrated microstructures are then numerically evaluated to compute physical properties such as ionic diffusivity or elastic modulus, to quantitatively relate microstructure to properties and design improved materials [4]. Initially, the microstructure model was based solely on the hydration of C_3S, the major component of portland cement, but it recently has been extended to include hydration reactions for all of the major phases [8].

The model is based on a cyclic process of dissolution, diffusion, and reaction, and is similar to a cellular automaton [8]. Basically, material dissolves in pixel-sized units from the cement particle surfaces, diffuses within the water-filled pore space, and reacts to form hydration products. The silicate phases form calcium silicate hydrate (C-S-H) gel and calcium hydroxide (CH) while the interstitial (aluminate and ferrite) phases form a variety of products (hydrogarnet- C_3AH_6, ettringite- AFt, and a monosulfoaluminate phase- AFm) depending on the amount of gypsum present in the system. Both surface-precipitated products such as the C-S-H gel and crystalline products such as CH are included in the rules representing the physical mechanisms of cement hydration. Both the chemical (molar) stoichiometry of the reactions that occur during hydration and the molar volumes of the products and reactants determine the volume (pixel) stoichiometry implemented within the digital-image-based computer model. For example, for each pixel of C_3S which dissolves, 1.75 pixels of C-S-H gel and 0.61 pixels of CH will be formed. Since the hydration products occupy a larger volume than the solid reactants, the cement paste ultimately converts from a viscous suspension into a rigid solid material.

It is well known that the various phases of cement react at different rates in a cement paste. Within the model, probabilities can be assigned to the dissolution processes so that the rank order of these phase reactivities can be maintained (e.g. $C_3A > C_3S > C_2S$, C_4AF). The mobility of various diffusing species may be controlled by choosing the location of a diffusing species relative to the location of the dissolution source. Silicate and iron diffusing species are located near the dissolution source to simulate their low mobility, while calcium, sulfate, and aluminate species are located at random throughout the microstructure, representing a uniform dispersion of these species. By monitoring the reactions occurring in a given cycle of hydration, the heat of hydration as a function of the number of cycles or the degree of hydration may be obtained. For this calculation, either the heats of formation of the cement compounds or tabulated values of the heats of hydration of the four main cement phases may be used [9,10].

Once cement particles have been stored in a database, they may be utilized to study the formation of interfacial zone microstructure. Interfacial zones in concrete have been shown to exhibit different microstructural characteristics than bulk cement paste both for cement paste-aggregate [11,12] and cement paste-steel rebar [13] interfaces. A simpler version of the NIST microstructure model has been successfully applied to simulating interfacial zone microstructures as a function of mineral admixture characteristics and aggregate absorptivity and reactivity [14]. For this model, a simple two-dimensional rectangular aggregate is first placed into the microstructure image and then the stored cement particles are added at random locations in the microstructure to obtain a representation of the desired global w/c ratio. The cement particles are not allowed to overlap any previously placed particle or any portion of the aggregate and are added in order of largest to smallest. The interfacial zone microstructure may be quantified by measuring the phase fractions as a function of distance from the aggregate surface both before and after hydration. These measurements may then be compared to those obtained from SEM images of real concrete specimens [14].

RESULTS

Area Fractions and Phase Perimeters

Table 1 shows the area and perimeter phase fractions for two different fields of view for the cement shown in Figure 3. It should be noted that the perimeter fractions will be a stronger function of image resolution than the area fractions. In both cases, the bulk area fractions are quite similar to the volume fractions computed using the Bogue equations [9]. Since most of the larger silicate particles are C_3S, the C_3S particles generally have a lower surface area to volume (or perimeter to area) ratio than the C_2S particles. Thus, the C_2S occupies a larger fraction of the phase perimeters than its area fraction. The C_2S generally reacts at a slower rate than the C_3S suggesting that this cement might hydrate at a slower rate than a cement with the same bulk phase fractions but a proportional distribution of the two silicates on its surfaces. Since the gypsum is generally present as smaller discrete particles, it too occupies a larger perimeter fraction than its area fraction.

Scrivener has computed the area and surface fractions of silicates and interstitial phases and in general found that the interstitial phases occupied a larger fraction of the surfaces than their area fraction [2]. For this particular cement, we actually observe the opposite trend, as the interstitial phases (ferrite and aluminate) occupy a smaller perimeter fraction than their area fraction. This can be directly observed in Figure 3, where many of the C_3A regions are part of much larger polymineralic particles and have much of their perimeter in contact with C_3S instead of porosity.

Table 1
Area and Perimeter Phase Fractions for a Type I Portland Cement

		Image 1		Image 2	
	Bogue Volume Fraction	Area Fraction	Perimeter Fraction	Area Fraction	Perimeter Fraction
C_3S	0.618	0.637	0.397	0.648	0.445
C_2S	0.127	0.115	0.322	0.121	0.316
C_3A	0.145	0.177	0.157	0.144	0.124
C_4AF	0.058	0.018	0.013	0.044	0.021
Gypsum	0.051	0.053	0.111	0.043	0.093

Heat of Hydration Results

Figure 4 shows two different initial cement particle images, both at a w/c of 0.45. The area and perimeter fractions for these two cement images are provided in Table 2. In Figure 5, calculated heat of hydration is plotted versus calculated degree of hydration for results obtained using the microstructure model to hydrate each cement for 200 cycles. As expected, a basically linear relationship is observed between heat evolved and degree of hydration. Since the two cements differ in bulk and surface compositions as shown in Table 2, their heat release characteristics also vary. Because Cement 1 has a higher C_3A content and C_3A has a higher heat of hydration than the other clinker phases [10], it releases more heat during hydration than Cement 2.

Interfacial Zone Microstructure

As mentioned previously, a database of stored cement particle shapes and phase distributions may be used along with the NIST microstructure model to simulate the development of microstructure in an interfacial zone in concrete. Figure 6 shows both the original random

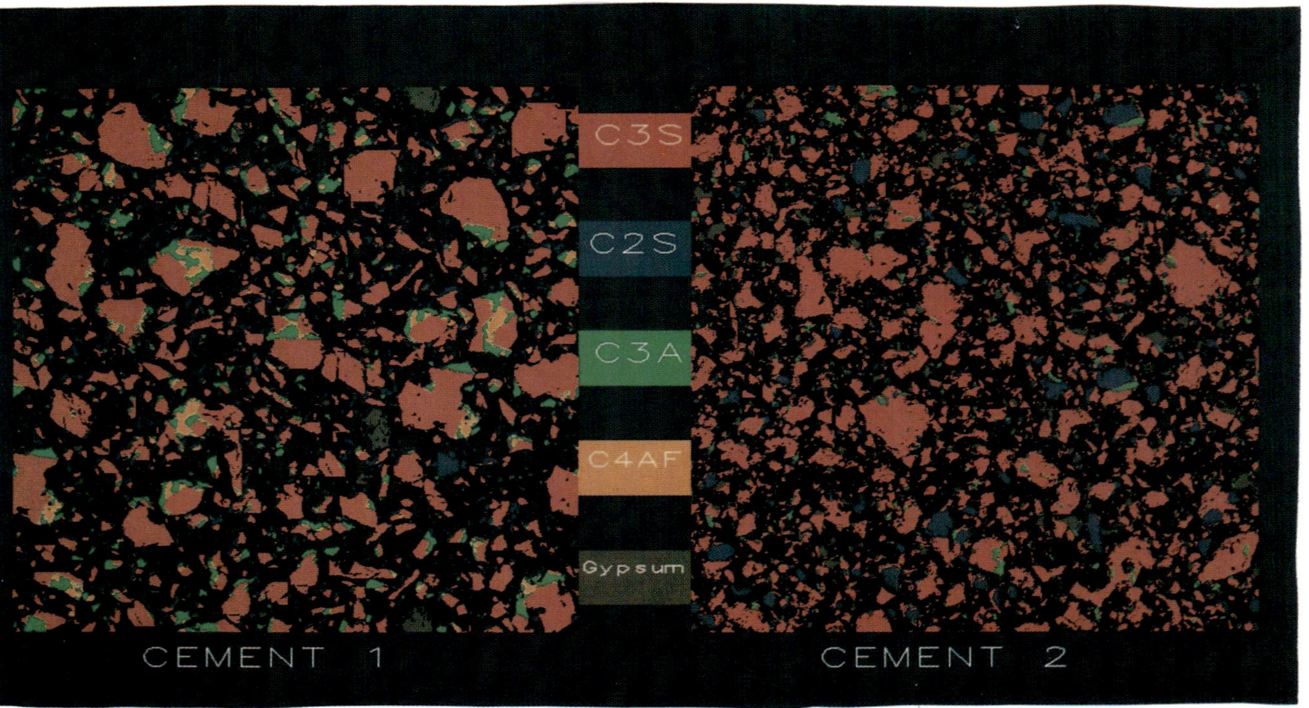

Figure 4. Computer-assembled 250*250 μm (500*500 pixel) images of initial cement particles for two cements, both with a w/c of 0.45. Color scheme is the same as in Figure 3. Cement 1 contains a higher fraction of C_3A (bright green) than Cement 2.

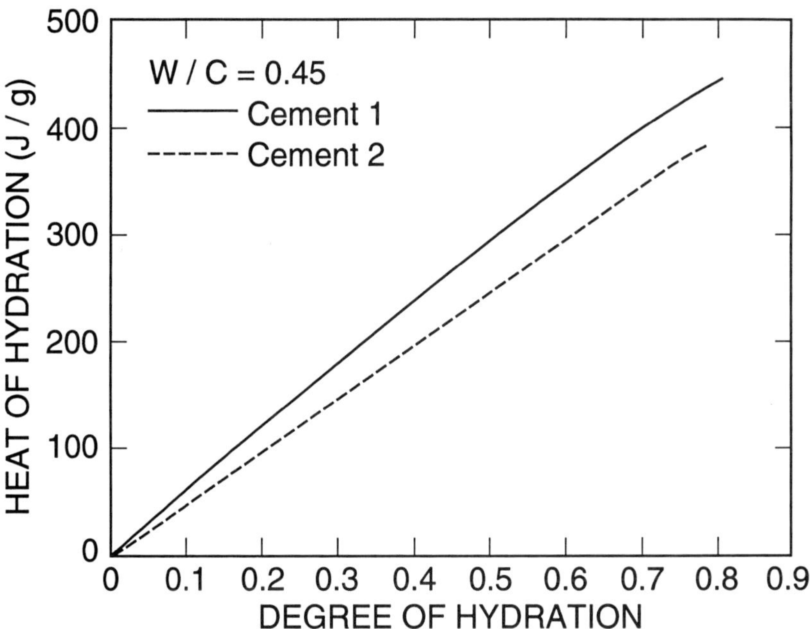

Figure 5. Heat of hydration curves calculated for the two cements shown in Figure 4.

Table 2
Phase Fractions for Cements in Figure 4

	Cement 1		Cement 2	
	Area Fraction	Perimeter Fraction	Area Fraction	Perimeter Fraction
C_3S	0.672	0.462	0.764	0.715
C_2S	0.118	0.307	0.110	0.101
C_3A	0.133	0.122	0.038	0.051
C_4AF	0.038	0.020	0.00	0.00
Gypsum	0.039	0.089	0.088	0.133

configuration of cement particles around an aggregate (purple) for a system with a w/c of 0.45 and the same system after hydration for 150 cycles using the microstructure model. In the hydrated image, C-S-H gel is yellow, CH and iron hydroxide (FH_3) are dark and bright blue respectively, and aluminate hydration products (AFt, C_3AH_6, etc.) are green. Since the cement particles do not pack efficiently in the vicinity of the aggregate, the interfacial or transition zone contains less cement and more water-filled porosity, thus having a higher local w/c ratio. This local inhomogeneity will affect both the hydration process and the resultant final microstructure.

The hydrated paste microstructure in Figure 6 appears to be different near the aggregate than the bulk paste microstructure far away

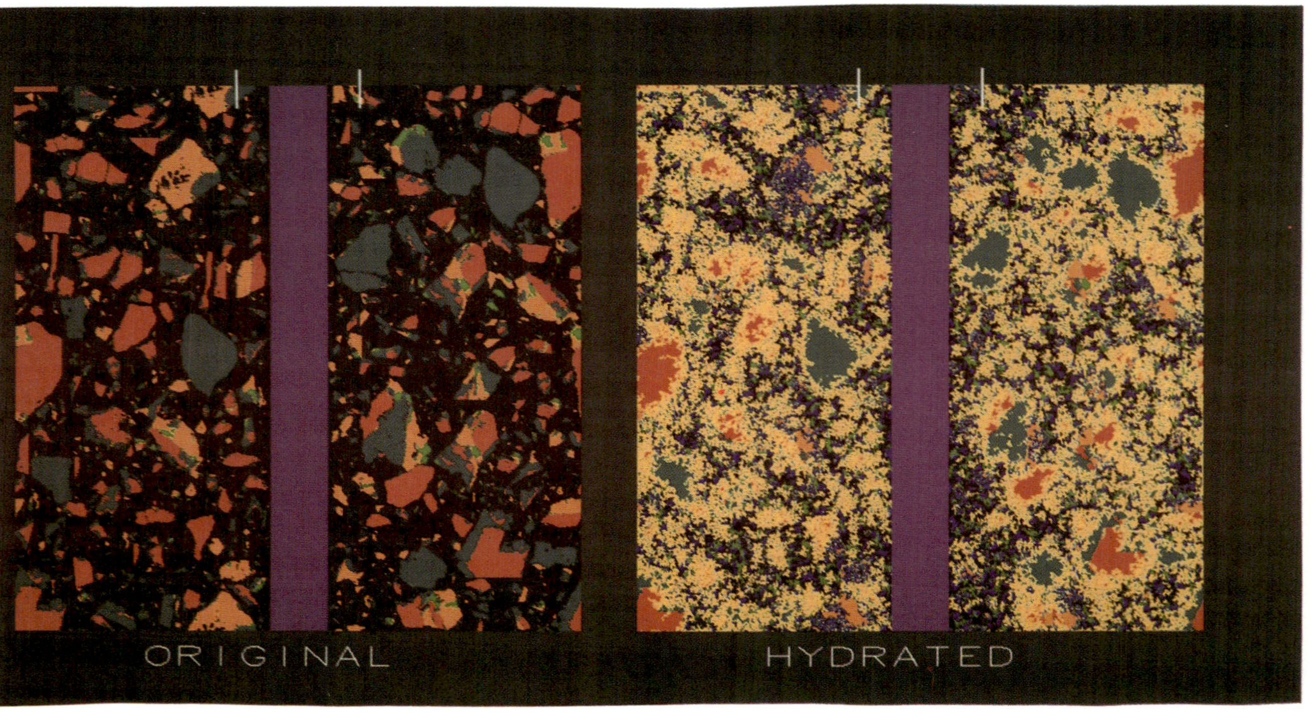

Figure 6. Original and hydrated images showing interfacial zone microstructure. Global w/c is 0.45. White bars at top of images indicate extent (30 pixels) of interfacial zones.

from the aggregate. To quantify this effect, Figure 7 shows a plot of the phase fractions as a function of distance from the aggregate surface for results averaged over five separate configurations of initial microstructures like the one shown in Figure 6. Even after 63% hydration, there is still a large increase in porosity as the aggregate surface is approached. In this region, there is also a deficiency of anhydrous cement, C-S-H, and FH_3. Conversely, the more mobile sulfate, calcium, and aluminate species lead to an increase in the CH, AFt, AFm, and C_3AH_6 in the area next to the aggregate. Since there is initially more porosity in this area and these mobile species tend to migrate throughout all available porosity, there is ultimately a larger volume of these hydration products formed near the aggregate than in the bulk paste. These computer model results are in agreement with numerous experimental observations [11-15].

X-ray Images of Hydrated Systems

Because the phase present at each pixel in an image is available as output from the microstructure model, simulated X-ray images can be easily produced. For example, a binary X-ray image for iron could be produced by highlighting all pixels where C_4AF or FH_3 is present. To demonstrate this process, Figure 8 shows binary X-ray images for calcium, silicon, iron, and aluminum for both the initial and hydrated interfacial zone model microstructures shown in Figure 6. In each case, the images to the left of the grey aggregate are the initial element distributions. Images on the right are mirror images of the ones on the left, but after 150 cycles of hydration. Here, it is clearly observed that the elements assigned a higher mobility in the model such as calcium and aluminum are indeed dispersed throughout the microstructure

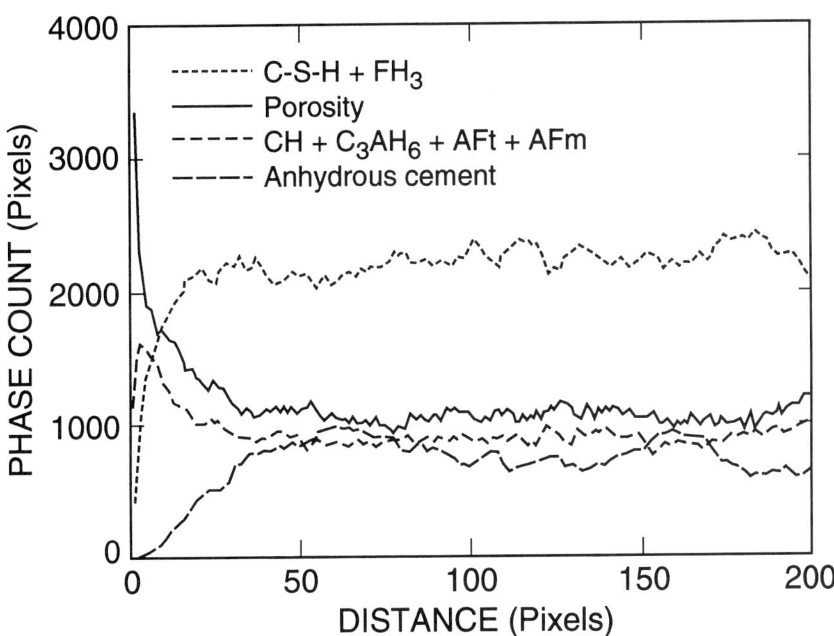

Figure 7. Phase distributions as functions of distance from aggregate surface for hydrated interfacial zone microstructure shown in Figure 6.

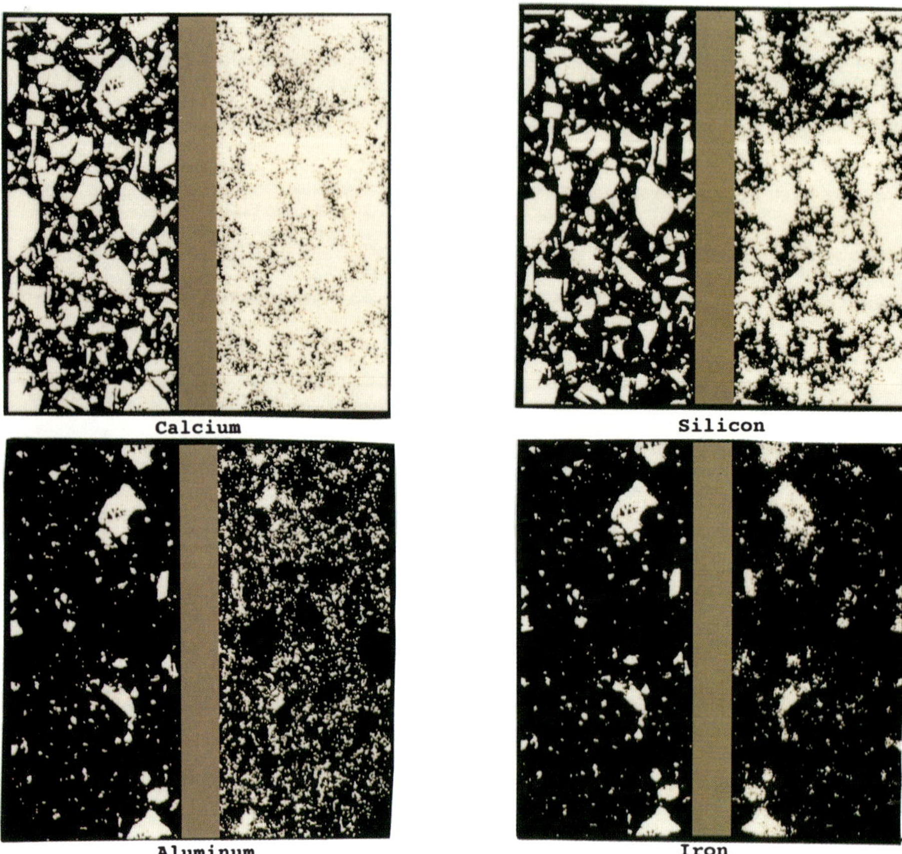

Figure 8. X-ray images for simulated interfacial zone microstructures from Figure 6, before (left) and after (right) 150 cycles (about 28 days) of hydration using the cement microstructure model.

during hydration, while the iron and silicon remain relatively close to their initial locations. These X-ray images could be compared to similar images obtained on real hydrated systems to validate and further improve the microstructure model.

CONCLUSIONS

A technique for obtaining two-dimensional images of cement particles segmented into individual phases has been demonstrated. The technique allows for quantitative characterization of cement powders, providing information on the distribution of phases not available from typical bulk oxide analyses. The technique can also be used to obtain starting images for cement paste microstructure computer models. The application of these computer models to predicting heat of hydration curves and studying interfacial zone microstructure in concrete has been demonstrated.

REFERENCES

[1] Stutzman, P.E., "Cement Clinker Characterization by Scanning Electron Microscopy", Cement, Concrete, and Aggregates, Vol. 13, No. 2, pp. 109-114, 1991.

[2] Scrivener, K.L., "The Microstructure of Anhydrous Cement and Its Effect on Hydration", in Microstructural Development During Hydration of Cement, Materials Research Society Symposia Proceedings Volume 85, Eds. L.J. Struble and P.W. Brown, pp. 39-46, 1987.

[3] Bonen, D., and Diamond, S., "Application of Image Analysis to a Comparison of Ball Mill and High Pressure Roller Mill Ground Cement", in Proceedings of the Thirteenth International Conference on Cement Microscopy, International Cement Microscopy Association, pp. 101-119, 1991.

[4] Bentz, D.P., and Garboczi, E.J., "Digital-Image-Based Computer Modelling of Cement-Based Materials", in Digital Image Processing: Techniques and Applications in Civil Engineering, Engineering Foundation Conference Proceedings, March 1993.

[5] Garboczi, E.J., and Bentz, D.P., "Computational Materials Science of Cement-Based Materials", MRS Bulletin, Vol. 18 (3), pp. 50-54, 1993.

[6] Castleman, K.R., Digital Image Processing, Prentice-Hall, Inc., Englewood Cliffs, NJ, 1979.

[7] Hagwood, C., "A Mathematical Treatment of The Spherical Stereology", NISTIR 4370, U.S. Department of Commerce, 1990.

[8] Bentz, D.P., Coveney, P.V., Garboczi, E.J., Kleyn, M.F., and Stutzman, P.E., "Cellular Automaton Simulations of Cement Hydration and Microstructure Development", submitted to Modelling and Simulation in Materials Science and Engineering.

[9] Taylor, H.F.W., Cement Chemistry, Academic Press, London, 1990.

[10] Neville, A.M., and Brooks, J.J., Concrete Technology, Longman Scientific and Technical, Essex, 1990.

[11] Scrivener, K.L., and Gartner, E.M., "Microstructural Gradients in Cement Paste Around Aggregate Particles", in Bonding in Cementitious Composites, Materials Research Society Symposium Proceedings, Eds. S. Mindess and S.P. Shah, pp. 77-85, 1988.

[12] Monteiro, P.J.M., and Mehta, P.K., "Ettringite Formation on the Aggregate-Cement Paste Interface", Cement and Concrete Research, Vol. 15, No. 2, pp. 378-380, 1985.

[13] Monteiro, P.J.M., Gjorv, O.E., and Mehta, P.K., "Effect of Condensed Silica Fume on the Steel-Cement Paste Transition Zone", Cement and Concrete Research, Vol. 19, No. 1, pp. 114-123, 1989.

[14] Bentz, D.P., Garboczi, E.J., and Stutzman, P.E., "Computer Modelling of the Interfacial Zone in Concrete", RILEM Proceedings Interfaces in Cementitious Composites, Ed. J.C. Maso, Vol. 18, pp. 107-116, 1992.

[15] Maso, J.C., "The Bond Between Aggregates and Hydrated Cement Pastes", Proceedings 7th International Cement Congress, pp. 3-15, 1980.

Paul E. Stutzman[1]

Applications of Scanning Electron Microscopy in Cement and Concrete Petrography

REFERENCE: Stutzman, P. E., **"Applications of Scanning Electron Microscopy in Cement and Concrete Petrography,"** Petrography of Cementitious Materials, ASTM STP 1215, Sharon M. DeHayes and David Stark, Eds., American Society for Testing and Materials, Philadelphia, 1994.

ABSTRACT: Due to its high resolution and analytic capabilities, scanning electron microscopy with image analysis provides several advantages in cement and concrete petrography. When combined with computer-based image analysis, scanning electron microscopy generates quantifiable images using its imaging capabilities of phase composition, its ability to image imaging element spatial distribution, and its chemical analysis capabilities. Applications of scanning electron microscopy in petrography of cementitious materials is illustrated by analyses of clinker, cement, and concrete microstructures, and in the characterization of concretes damaged by alkali-silica and sulfate attack.

KEYWORDS: cement, characterization, clinker, concrete, image analysis, microstructure, scanning electron microscopy, X-ray microanalysis.

Methods for analysis of concrete-making materials are being developed at the Building and Fire Research Laboratory of the National Institute of Standards and Technology. These methods include application of petrographic techniques to characterize clinker, cement, aggregate, and other concrete-making materials. New or improved methods for characterizing materials will provide information necessary for evaluating the durability of concrete and for the prediction of service life. Scanning electron microscopy has an important role in the characterization of the microstructures of these materials. Cement clinker, cement, hardened cement paste, and concrete can be studied in the scanning electron microscope (SEM) in a manner similar to optical microscopy (OM). While there are overlaps in capabilities, the SEM has a number of distinct advantages in imaging and measurement, including A) the ability to produce high resolution images with a relatively large depth of field; B) sufficient contrast in imaging polished sections to readily study phase abundance and distribution; C) imaging element spatial distribution, D) quantitative description of the images via image processing and analysis; and E) quantitative chemical analysis.

This paper presents a review of the SEM, provides examples of backscattered electron and X-ray imaging techniques, and chemical analyses, illustrates the application of image processing and analysis to quantify microstructural features, and applies these techniques to the evaluation of cement and concrete.

---

[1]Physical Scientist, Building Materials Division, Building and Fire Research Laboratory, National Institute of Standards and Technology, U.S. Department of Commerce, Gaithersburg, MD 20899.

Preparation of Samples for SEM Analysis

Careful sample preparation is critical for the study of cement paste and concrete. Dry specimens are placed in a vacuum oven at 60°C to remove any residual moisture, then vacuum-impregnated with a low-viscosity resin, which is cured at 60°C. Shrinkage cracking during drying is a problem with the preparation of cement and concrete specimens. Therefore an alternative procedure has been developed at NIST [1] that minimizes cracking due to drying shrinkage by allowing replacement of the pore solution with resin without drying the specimen.

Fine polishing is done with a lapidary wheel covered with lint-free cloth, using diamond paste (6, 3, 1, and 0.25 μm) as an abrasive. A properly polished specimen should contain flat, scratch-free unhydrated cement particles with a well-defined internal structure, and calcium hydroxide and the capillary pores should appear distinct. Samples are electrically insulating and need to be coated (about 15 nm thick) with a conductive material such as carbon.

SEM Imaging

The SEM produces and scans a finely-focused beam of electrons across the specimen and measures signals resulting from the electron beam/specimen interaction. Common signals employed in SEM analysis include secondary electrons for imaging surface topography; backscattered electrons for highlighting compositional differences; and X-rays for determining elemental composition and imaging element spatial distribution.

Backscattered electron (BE) imaging is an ideal technique for distinguishing phases in a microstructure when using flat, polished surfaces and is commonly referred to as compositional imaging. Image contrast is generated by different phases' compositions relative to their weighted average atomic number ($\bar{Z}$). BE signal intensity, the backscattered electron coefficient (η), is the ratio of the number of backscattered electrons to that in the incident beam at the specimen surface (13). Higher $\bar{Z}$ phases appear brighter (Table 1). The polished surface necessary for compositional BE imaging is also ideally suited for quantitative analysis of microstructural features. Images are monochrome since they reflect the electron or X-ray flux resulting from the beam/specimen interaction. Backscattered electron and X-ray imaging are the most useful imaging modes for quantitative microscopy and will be the focus of this paper.

Table 1-- Common clinker and cement phases, average atomic number ($\bar{Z}$) and backscattered electron coefficient (η)*.

PHASE	$\bar{Z}$	η	PHASE	$\bar{Z}$	η
ALITE	15.06	.176	CALCIUM SILICATE HYDRATE	13.68	.161
BELITE	14.56	.171	CALCIUM HYDROXIDE	14.30	.168
ALUMINATE	14.11	.166	MONOSULFATE	11.66	.137
FERRITE	17.12	.198	ETTRINGITE	10.79	.127
FREE LIME	16.58	.193	HYDROGARNET	13.93	.164
PERICLASE	10.41	.122	THAUMASITE	9.74	.114
ARCANITE	14.41	.169	QUARTZ	10.81	.127
GYPSUM	12.12	.143	CALCITE	12.57	.158
ANHYDRITE	13.42	.158	DOLOMITE	10.87	.128

*$\bar{Z}$ and η values vary with composition; brightness will be influenced by porosity.

X-Ray Microanalysis

X-radiation is produced when a specimen is bombarded by high energy electrons. The X-ray energy level is displayed as the number of counts at each energy level and plotted as a set of peaks on a continuous background. The positions of the peaks are characteristic of a particular element, so identification is made by peak positions and relative intensities. The X-ray signal can be used in a number of ways: A) spectrum analysis to determine which elements are present and in what concentration; B) line scan analysis to display the relative concentration changes along a line; C) X-ray imaging (XR) of element spatial distribution and relative concentrations, and D) as an aid in phase identification. Mass concentration to a few tenths of a percent can be detected using an energy dispersive X-ray detector. Relative accuracy of quantitative analysis is about ±20% for concentrations of about 1%, and ±2% for concentrations greater than 50% [2].

Magnification and Resolution

Magnification and resolution are important concepts in microscopy. Magnification is the ratio of a line on a photomicrograph to a similar line on the specimen. A scale bar is preferred over magnification since the size of a photomicrograph may vary upon reproduction.

Resolving power is the ability to distinguish close, separate objects. The unaided human eye can resolve points about 120 μm apart, while the white-light optical microscope with an oil immersion lens is capable of resolving points 0.2 μm (200 nm) apart [3]. SEM resolution depends on the operating conditions, imaging mode, and the nature of the specimen. While SE image resolution is 10 nm or better, BE images of cement paste and cement clinker, at 12 kV accelerating voltage, should attain a resolution of about 0.4 μm, a size corresponding to the coarse capillary porosity. While BE image resolution does not exceed that of OM, image contrast allows better definition of the constituents. X-ray resolution is slightly less than 1 μm.

Analysis of Images

The application of computers for image analysis has strengthened the utility of the SEM in quantitative microscopy by providing a means to collect, process, and analyze images. Image processing encompasses techniques used to isolate features of interest. Image analysis quantifies those features by counting particles, describing particle shapes, determining phase fractions, and establishing relationships between elements and phases.

However, for any microscopy technique, increased magnification reduces the sampling area and representative sampling may become a problem. Using an example by Diamond [4], ten BE images at 750x magnification from a 1 cm square specimen samples 0.2% of that specimen's microstructure. Computer-based image analysis minimizes difficulties in collecting and performing measurements on large numbers of images.

Microstructure of Clinker and Cement

Microscopy analysis of clinker and cement provides information not only on which phases are present, but also on their abundance, distribution, crystal form and size, and porosity. Optical microscopy has traditionally been used to study clinker and cement [5,6]. However, etching used in OM to aid phase identification produces sample relief often greater than the depth of focus. For cements, etching may dissolve the finer-sized particles. Examination of polished sections of clinker and cement particles can be similarly accomplished using SEM imaging and surface etching is not required.

Scrivener [7] determined phase fractions and perimeter phase fractions of anhydrous cement grains by analysis of BE images. Stutzman et. al [8] and Stutzman [9] used SEM imaging to describe the NIST Reference Material clinkers. Diamond and Olek [10] analyzed BE and XR images to characterize cement particle shape and composition. Bonen and Diamond [11] used image analysis to examine the differences in cements ground in ball mills and high-pressure roller mills. Bentz and Stutzman [12] imaged cements to determine phase abundance, distribution, and perimeter phase fraction.

A BE image of a clinker fragment is shown in Figure 1. Criteria for phase identification include relative brightness (Table 1), crystal morphology, association, and chemical composition (Fig. 2). By association for example, a phase could be either part of the interstitial material, a framework grain, or occur infrequently throughout the microstructure. XR imaging aids in location of some of the infrequent phases such as periclase and the alkali sulfates.

X-ray peak intensity is proportional to the concentration of an element. Reference standards and corrections for inter-element effects are used in making quantitative analyses. The method used here (ZAF) calculates the peak intensity ratio of the specimen and standard, and then corrects for inter-element effects based on atomic number (Z), absorption (A), and fluorescence (F) [13]. Quantitative compositional analyses for some clinker phases of Reference Material (RM) 8488 are given in Table 2.

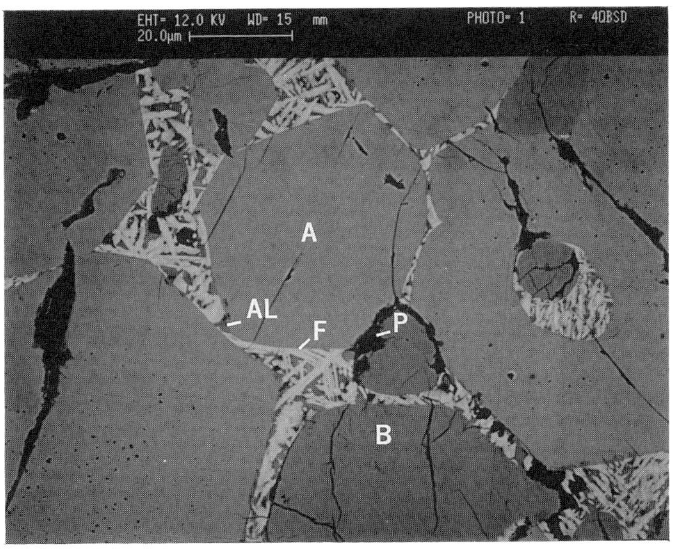

Figure 1. BE image of RM 8488 with coarsely-crystalline alite (A) and belite (B), dendritic to lath-like ferrite (F), fine-grained aluminate (AL), and porosity (P).

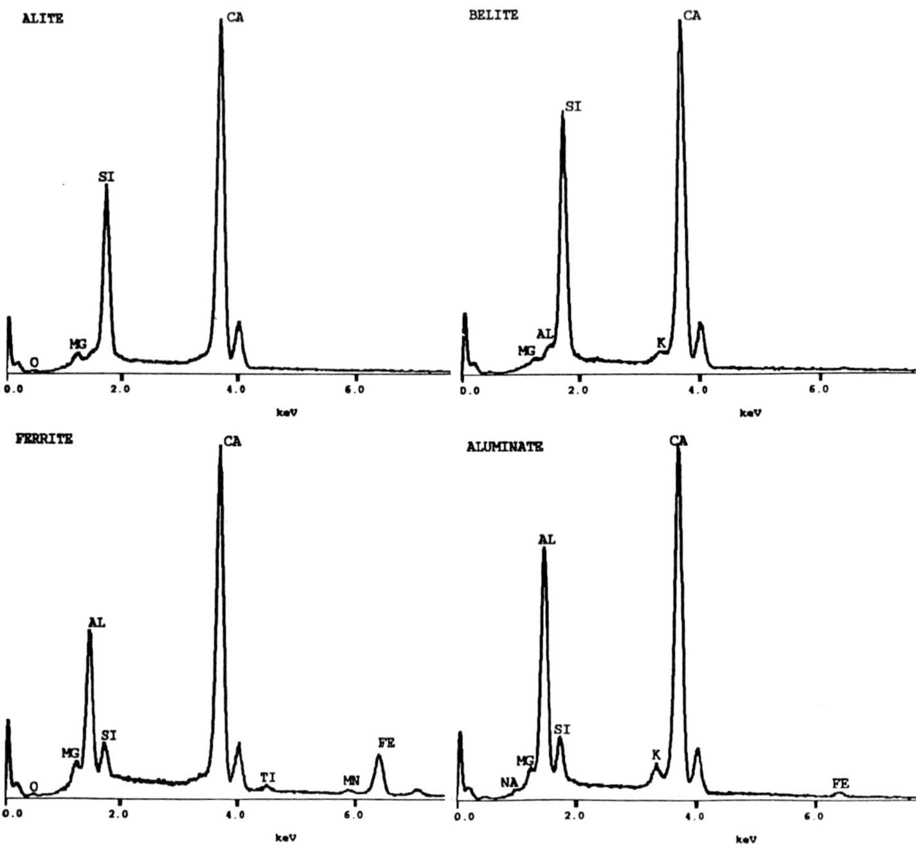

Figure 2. Energy-dispersive X-ray (EDS) spectra are used for qualitative and quantitative compositional analysis.

Table 2-- Compositional analysis of selected clinker phases of RM 8488 expressed as oxides.

PHASE	CaO	SiO_2	Al_2O_3	MgO	Fe_2O_3	K_2O	Na_2O	TiO_2	Mn_2O_3
ALITE	72.6	25.1	0.7	1.6	0.0	0.0	0.0	0.0	0.0
BELITE	64.9	31.9	1.0	0.5	1.0	0.7	0.0	0.0	0.0
ALUMINATE	57.7	4.3	31.7	0.3	3.6	1.6	0.8	0.0	0.0
FERRITE	49.3	4.1	20.4	2.6	21.6	0.0	0.0	1.0	1.0

Cement imaging is particularly difficult because of the fine particle sizes. Point count analysis by X-ray spectrum matching has shown promise as an automated procedure for determining phase abundances. For this procedure, reference spectra are collected for each phase, then X-ray spectra from a grid of points from multiple fields are collected, and finally a matching routine finds the best-fit reference spectrum for each unknown. SEM/EDS point count data of the Reference Material clinkers compares very well to certificate values determined by optical microscopy (Fig. 3). Measurement error in quantitative microscopy analyses has been discussed by Galehouse [14] and Campbell [5]. For point count analysis, the half width of an approximately 95 percent confidence interval can be calculated using (1):

$$E_{95.4} = 2\sqrt{\frac{P(100-P)}{N}} \quad (1)$$

Where E = the half width in percent
N = the total number of points counted, and
P = the percentage of the phase.

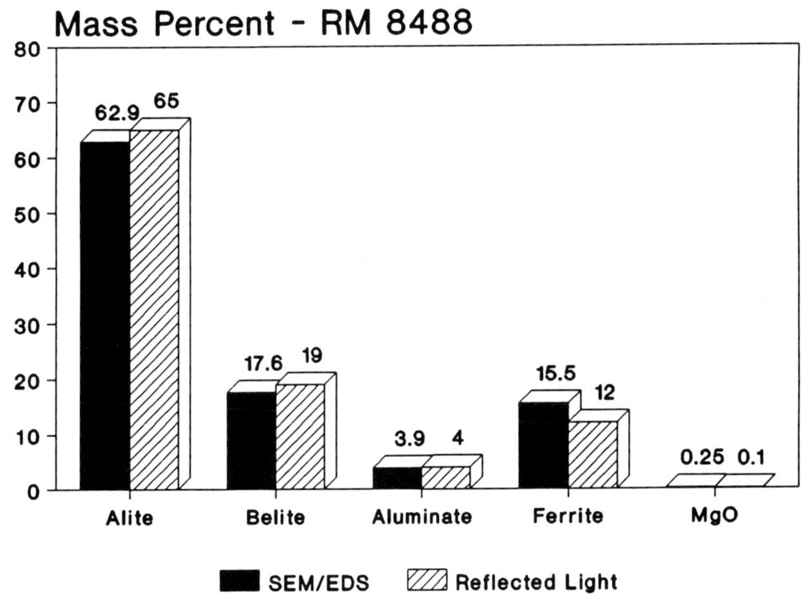

Figure 3. Automated SEM/EDS point count of RM 8488 clinker phase abundances. Certificate values were determined using reflected light microscopy. Half-width values with 1400 points counted are alite, ±2.6%; belite, ±2.0%; aluminate, ±1.0%; ferrite, ±1.9%; and periclase, ±0.27%.

Microstructure of Cement Paste in Concrete

SEM and image analysis have been used to study the microstructure of cement paste and the paste/aggregate interfacial zone [15,16], microstructural development at different curing temperatures [17,18], and pore structure [19].

Figure 4 shows the microstructure of a paste/aggregate interfacial zone in a 24-hour-old, 0.45 water/cement (w/c) concrete specimen. The brightest particles are of unhydrated cement (AN). Calcium hydroxide (CH) can occur in any of three morphologies: elongated crystals exhibiting distinct basal cleavage; blocky masses; and finely disseminated crystals not resolvable by BE imaging. Two morphologically distinct forms of calcium silicate hydrate (CSH) are seen in hardened cement paste. The massive, dense-appearing CSH that envelopes unhydrated cement grains and replaces smaller cement grains is thought to form in situ and is termed 'inner product' or 'late product'. The CSH formed in the water-filled void space appears darker with visible porosity and is termed 'outer product' or 'undesignated product' [20]. Taylor and Newbury [21] found that the two CSH forms do not differ significantly in composition, but that the undesignated product may contain intermixed, sub-micron particles of other phases.

The grey-level histogram for the image in Fig. 4 is a graphical representation of the number of pixels in each of 256 levels of grey ranging from black to white. The BE image can be processed to segment the image based on grey-levels and then analyzed to determine phase fractions.

In comparison with Fig. 4, Figs. 5a and 5b show distinct changes in the concrete paste microstructure after 7 days of curing. X-ray images show the spatial distribution of selected elements with image brightness proportional to X-ray intensity. X-ray imaging simplifies location of phases not apparent in the BE image. Calcium and silicon distributions are fairly uniform, brighter calcium regions correspond to CH and AN and brighter silicon regions correspond to AN.

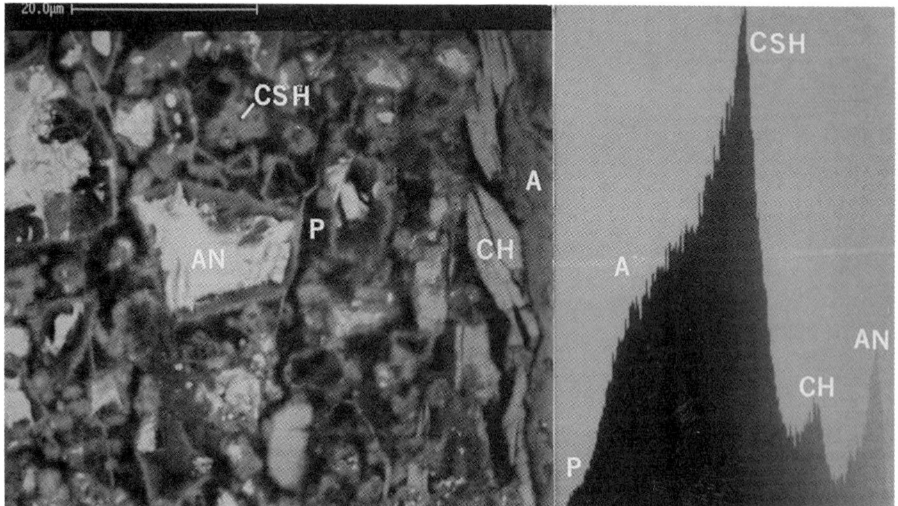

Figure 4. BE image and grey-level histogram of a 24-hour-old 0.45 w/c concrete with unhydrated cement (AN), calcium hydroxide (CH), calcium-silicate-hydrate (CSH), aggregate (A), and porosity (P).

Aluminum distribution is less uniform, with bright regions corresponding to aluminate and ferrite. High-aluminum regions which also exhibit relatively high sulfur are either monosulfate or ettringite. Comparing X-ray spectra to reference spectra should help identification. Monosulfate appears as a relatively dark, uniform phase, sometimes exhibiting a platy parting.

Magnesium and iron have a low solubility in an alkaline environment and tend not to migrate far from their origin [21]. Bright areas in the magnesium image correspond to unhydrated periclase in the cement and hardened paste. Some high-magnesium regions may be $Mg(OH)_2$, since these particles exhibit a parting not present in periclase. Outlines of the original cement particles can be seen in the weaker regions of the magnesium image. Estimates of the original water/cement ratio may be made by reconstructing the original cement grains and measuring the area fractions of the cement and void space. Iron appears to be primarily associated with unhydrated interstitial material.

Potassium distribution in the hardened paste shows a strong correlation with sulfur with these regions appearing brighter yet morphologically similar to the undesignated product. This phase may be syngenite ($K_2Ca(SO_4)_2 \cdot H_2O$), or material precipitated from the pore solution [21].

Using the original mix design data, the degree of hydration can be calculated by measuring the amount of residual unhydrated cement using image analysis [18]. The initial cement, mineral admixtures, and water contents are converted to volume fractions by dividing the mass fraction by its density and the degree of hydration (α) calculated using (2):

$$\alpha = 1 - \frac{ResidualCement}{InitialCement} \qquad (2)$$

Microstructural differences between the bulk cement paste and cement paste near the aggregate can also be characterized by SEM imaging. The transition zone can approach 100μm depending on the water/cement ratio, and is characterized by higher porosity, greater CH abundance, and lower CSH and AN abundance than the bulk paste. Besides being a zone of weakness, the transition zone's greater permeability may influence concrete durability.

Mineral Admixtures

Fly ash and blast furnace slag (Fig. 6) can be observed using BE and XR imaging. Silica fume has a pronounced affect on the microstructure of the paste and transition zone (Fig. 7a). Bentz et. al [22] found that with greater amounts of silica fume, the CH content decreases, the overall microstructure becomes much more uniform, and the transition zone decreases. These changes are generally attributed to the pozzolanic reaction between CH and silica fume and an improvement in packing density due to the fine particle size of silica fume. Image analysis of the transition zone of specimens with 0, 10, and 20 percent silica fume are presented in Fig. 7b.

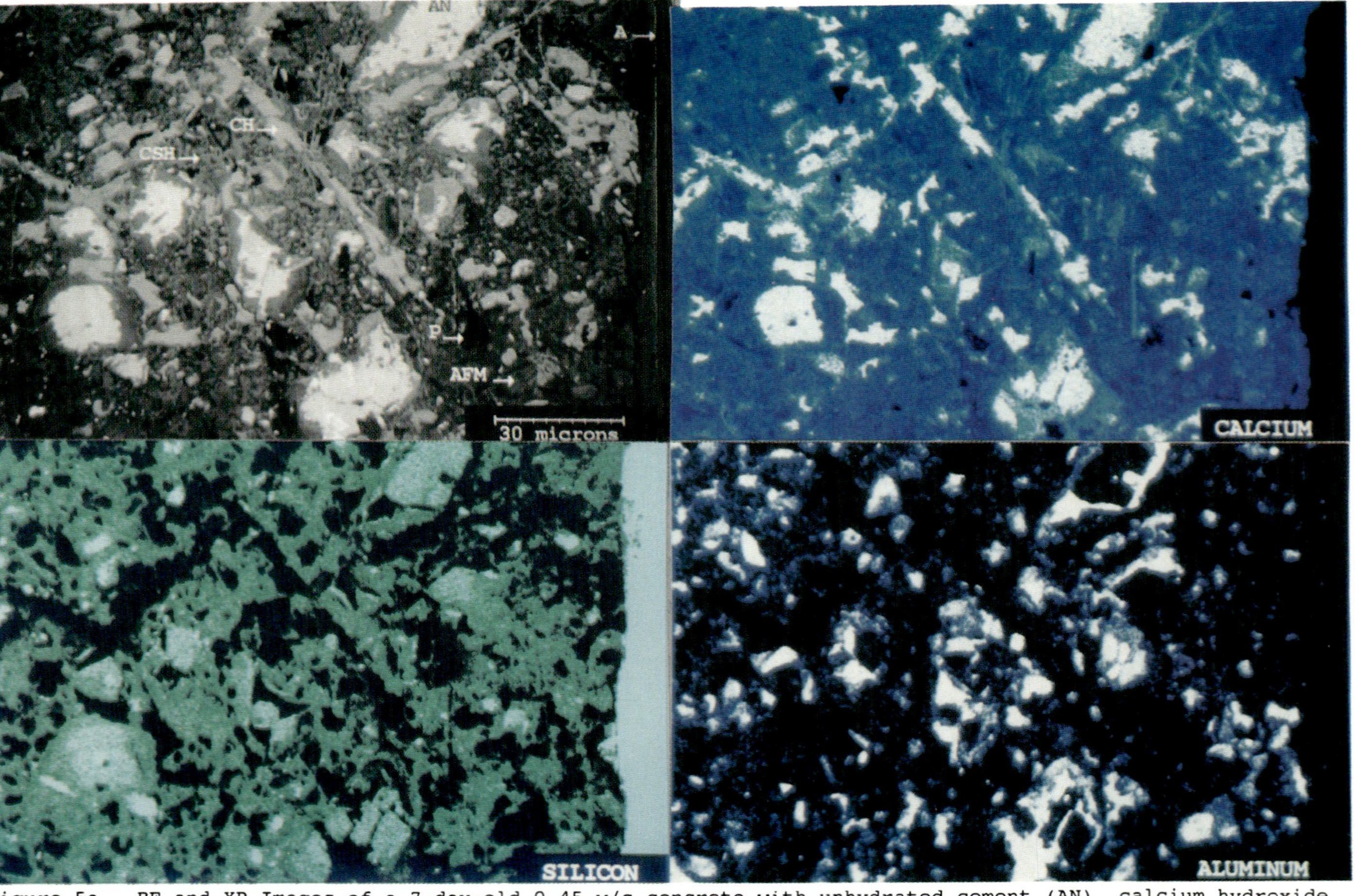

Figure 5a. BE and XR Images of a 7-day-old 0.45 w/c concrete with unhydrated cement (AN), calcium hydroxide (CH), calcium-silicate-hydrate (CSH), monosulfate (AFM), periclase (M), aggregate (A), and porosity (P).

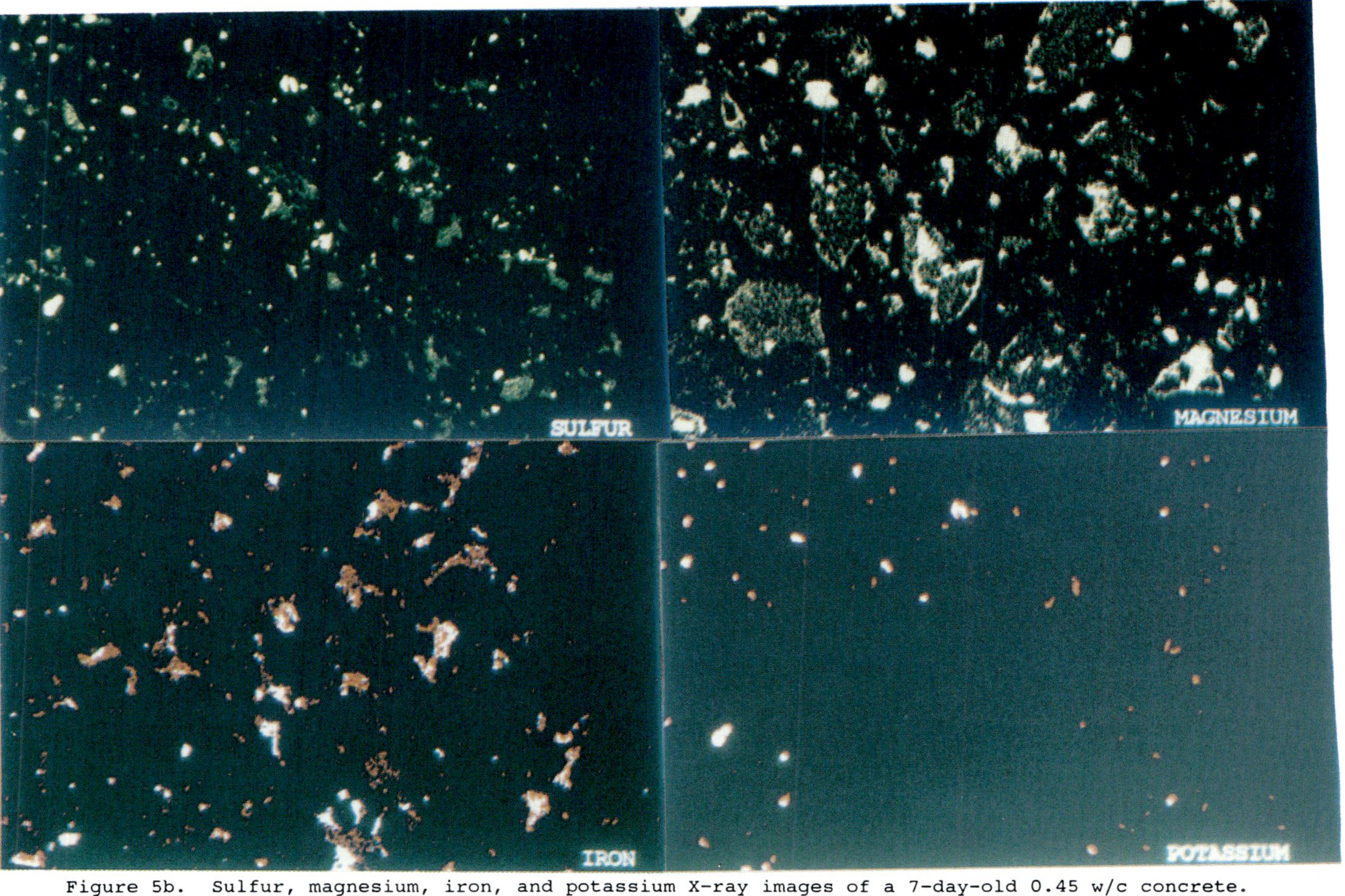

Figure 5b. Sulfur, magnesium, iron, and potassium X-ray images of a 7-day-old 0.45 w/c concrete.

(a)

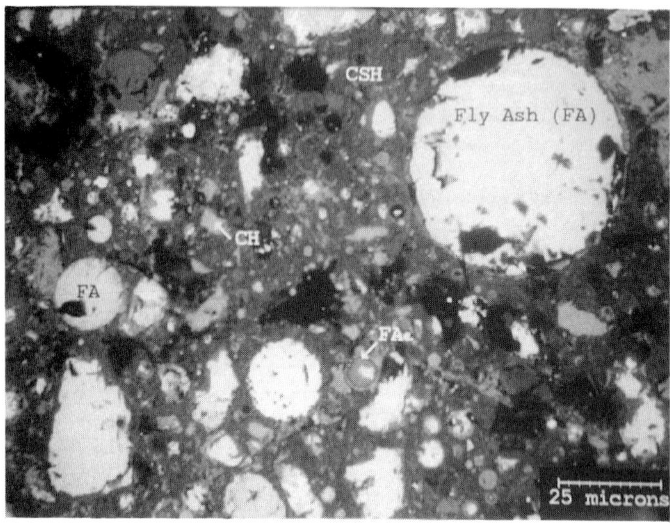

(b)

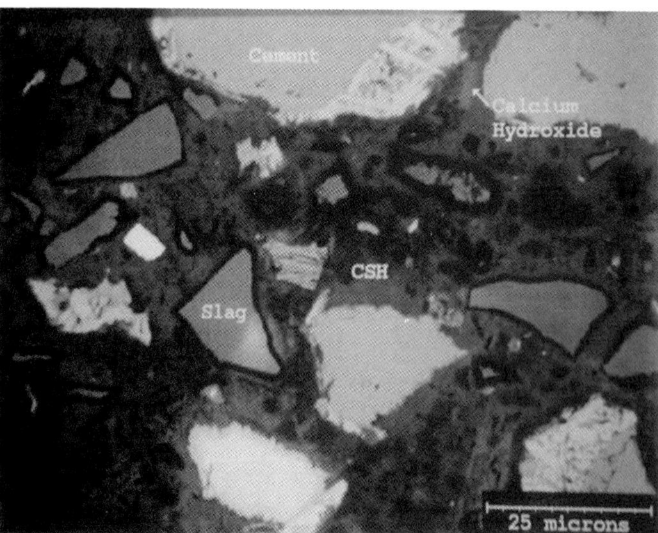

Figure 6. Mineral admixtures such as fly ash (a) and ground blast furnace slag (b) can also be imaged by SEM.

(a)

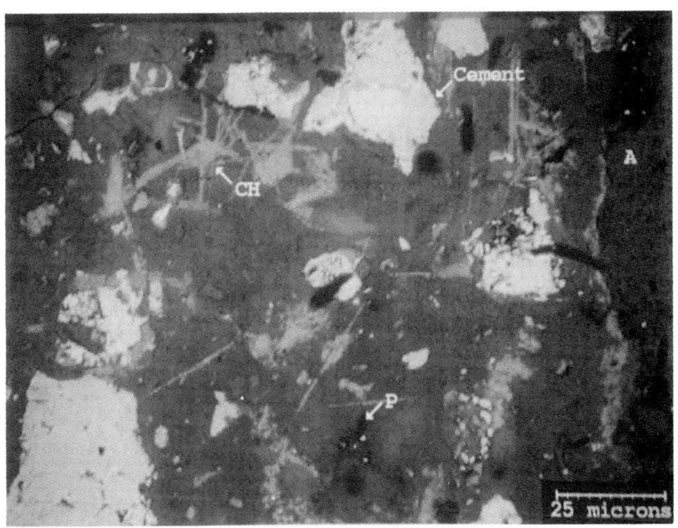

(b)

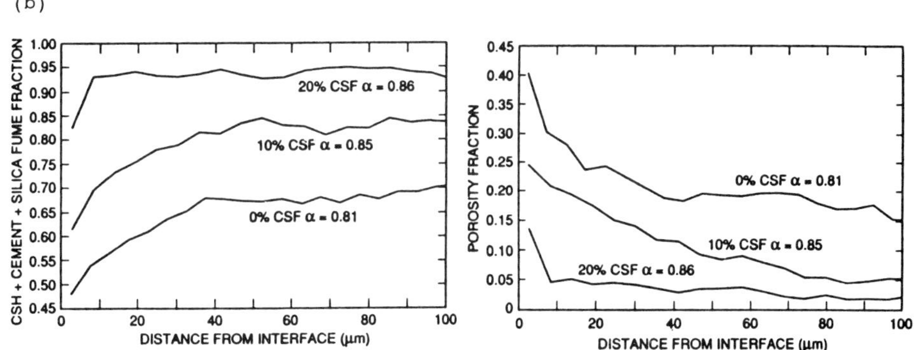

Figure 7. Cement paste microstructure of a 7-day-old 0.45 water/solids concrete with 20% condensed silica fume (CSF) replacement of cement (a) and graphical representation of the change in phase fractions from BE images illustrating the influence of silica fume on concrete microstructure (b) [22].

Chemical Attack in Field Concrete

A slab extracted from a core of a sidewalk that exhibited fine map-cracking was examined by BE and XR imaging (Fig. 8). X-ray analysis of small pockets of reaction gel indicated the presence of potassium, sodium, calcium, and silicon. Examination of X-ray images of potassium and sodium distribution indicated additional reaction gel occurrences near and within the aggregate. Since the gel composition changes outside the aggregate, the dense uniform texture and occasional cracking can be used for identification. Lack of entrained air may also be contributing to the deterioration of this sidewalk.

A specimen of extremely friable concrete from a semi-arid Western environment was examined by X-ray powder diffraction, BE, and XR imaging. The powder diffraction analysis indicated large amounts of ettringite and gypsum. SEM images (Fig. 9) show an abundance of an acicular material with an intermediate sulfur intensity and some pockets of a brighter, massive phase with high sulfur intensity along the paste/aggregate interfacial zone. X-ray spectrum matching identified the acicular material as ettringite and the massive phase as gypsum.

Summary

The value of the SEM in cement and concrete petrography is in its ability to image and make measurements on microstructural features, and in qualitative and quantitative chemical analysis. Backscattered electron and X-ray imaging of polished surfaces yield images with sufficient contrast that phases can be observed. Computer-based image analysis facilitates microstructure characterization. X-ray analysis can be used for phase identification, and qualitative and quantitative chemical analysis. These capabilities make the SEM a valuable addition to the petrography laboratory.

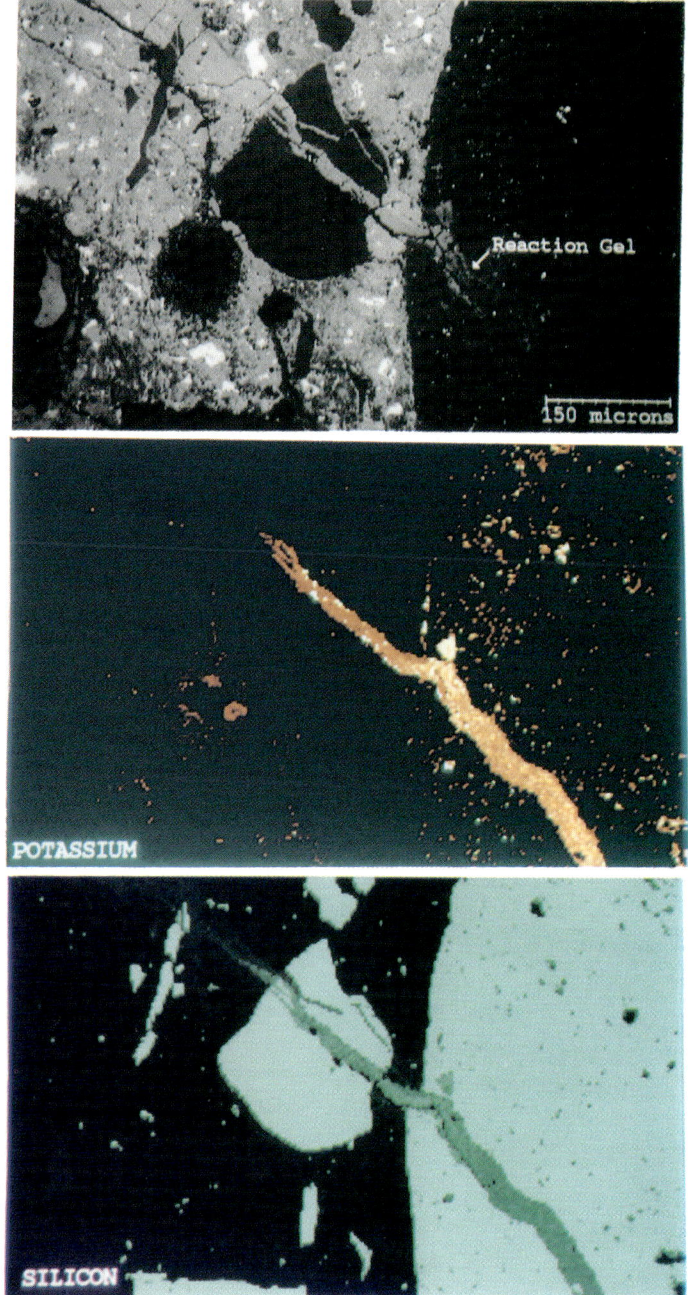

Figure 8. Backscattered electron and X-ray images of potassium and silicon distribution are used to locate alkali-silica reaction gel in concrete.

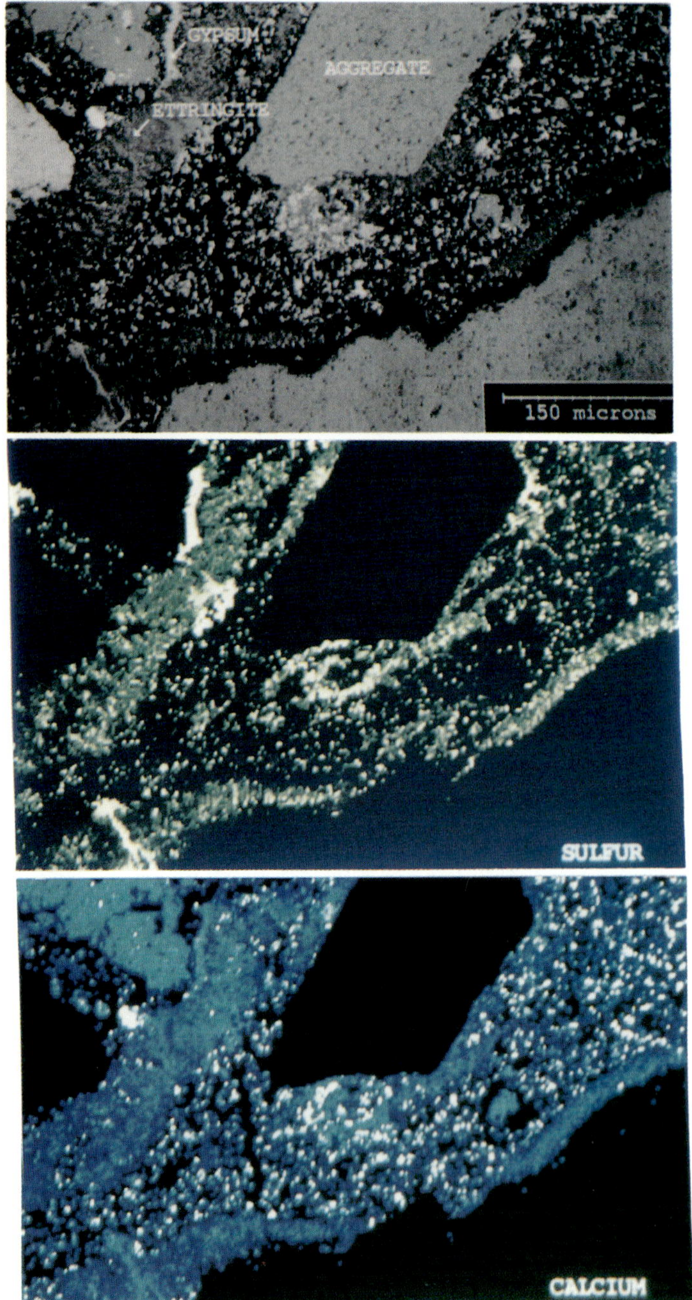

Figure 9. Deposits of ettringite and gypsum in a concrete damaged by sulfate attack are identified by BE and XR imaging along with X-ray spectrum matching.

Acknowledgements

The author acknowledges with appreciation the support of the Building Materials Division, the review and suggestions of Jim Clifton, Jim Pommersheim, Jim Pielert, and Steve Lane, and the editorial suggestions of Marcia Stutzman.

REFERENCES

[1] Friel, J., "X-Ray Microanalysis by Energy Dispersive Spectroscopy", Standardization News, Vol. 19, No. 5, pp. 38-41, 1991.

[2] Struble, L.J. and Stutzman, P.E., "Epoxy Impregnation of Hardened Cement for Microstructural Characterization", Journal of Materials Science Letters (8), 1989, pp. 632-634.

[3] McCrone, W. C., McCrone, L.B., and Delly, J.G., Polarized Light Microscopy, McCrone Research Institute, 1984.

[4] Diamond, S., "The Microstructures of Cement Paste in Concrete", Proc. 8th Intl. Cement Congress, pp. 123-147, 1986.

[5] Campbell, D.H., Microscopical Examination and Interpretation of Portland Cement and Clinker, Portland Cement Association, 1986.

[6] Campbell, D.H. and Galehouse, J.S., "Quantitative Clinker Microscopy with the Light Microscope", Cement, Concrete, and Aggregates, Vol. 13, No. 2, pp. 94-96, 1991.

[7] Scrivener, K. L., "The Microstructure of Anhydrous Cement and its Effect on Hydration", in Microstructural Development During Hydration of Cement, Materials Research Society Symposia Proceedings, Eds. L.J. Struble and P.W. Brown, pp. 39-46, 1987.

[8] Stutzman, P.E., Lenker, S., Kanare, H., Tang, F., Campbell, D., and Struble, L., "Standard Cement Clinkers for Phase Analysis", Proc. of the 11th Internat. Conf. on Cement Microscopy, pp. 154-168, 1989.

[9] Stutzman, P.E., "Cement Clinker Characterization by Scanning Electron Microscopy", Cement, Concrete, and Aggregates, Vol. 13, No. 2, pp. 109-114, 1991.

[10] Diamond, S. and Olek, J., "Cement Particle Characterization by SEM - Chemical Image Analysis Systems", Proc. of the 12th Internat. Conf. on Cement Microscopy, pp. 356-369, 1990.

[11] Bonen, D. and Diamond, S., "Application of Image Analysis to a Comparison of Ball Mill and High Pressure Roller Mill Ground Cement", Proc. 13th Internat. Conf. on Cement Microscopy, pp. 101-119, 1991.

[12] Bentz, D.P. and Stutzman, P.E., "SEM Analysis and Computer Modelling of Hydration of Portland Cement Particles", submitted to ASTM STP 1215, Petrography of Cementitious Materials, 1993.

[13] Goldstein, J.I., Newbury, D.E., Echlin, P., Joy, D.C., Fiori, C., and Lifshin, E., Scanning Electron Microscopy and X-Ray Microanalysis, Plenum Press, New York, 1981.

[14] Galehouse, J.S., "Point Counting" in Carver, R.E., Ed., Procedures in Sedimentary Petrology, Wiley, New York, 1971.

[15] Scrivener, K.L., "Quantification of Microstructure", in Advances in Cement Manufacture and Use, Proc. of the Engineering Foundation Conf., E. Gartner, Ed., pp. 3-13, 1988.

[16] Scrivener, K.L. and Gartner, E.M., "Microstructural Gradients in Cement Paste Around Aggregate Particles", in Bonding in Cementitious Composites, Materials Research Society Symposium Proceedings, Eds. S. Mindess and S.P. Shah, pp. 77-85, 1988.

[17] Kjellsen, K.O., Detwiler, R.J., and Gjørv, O.E., "Backscattered Electron Imaging of Cement Pastes Hydrated at Different Temperatures", Cement and Concrete Research, Vol. 20, pp. 308-311, 1990.

[18] Stutzman, P.E. and Clifton, J.R., "Microstructural Features of Some Low Water/Solids Silica Fume Mortars Cured at Different Temperatures", NISTIR 4790, 1992.

[19] Scrivener, K.L., "The Use of Backscattered Electron Microscopy and Image Analysis to Study the Porosity of Cement Paste", in Pore Structure and Permeability of Cementitious Materials, Eds. L.R. Roberts and J.P. Skalny, pp. 129-140, 1989.

[20] Taylor, H.F.W., Cement Chemistry, Academic Press, London, 1990.

[21] Taylor, H.F.W. and Newbury, D.E., "An Electron Microprobe Study of a Mature Cement Paste", Cement and Concrete Research, Vol. 14, No. 4, pp. 656-573, 1984.

[22] Bentz, D.P., Stutzman, P.E., and Garboczi, E.J., "Experimental and Simulation Studies of the Interfacial Zone in Concrete", Cement and Concrete Research, Vol. 22, pp. 891-902, 1992.

S. Wirgot[1] and F. Van Cauwelaert[2]

THE INFLUENCE OF CEMENT TYPE AND DEGREE OF HYDRATION ON THE MEASUREMENT OF W/C RATIO ON CONCRETE FLUORESCENT THIN SECTIONS

REFERENCE: Wirgot, S. and Van Cauwelaert, F., "**The Influence of Cement Type and Degree of Hydration on the Measurement of W/C Ratio on Concrete Fluorescent Thin Sections**," Petrography of Cementitious Materials, ASTM STP 1215, Sharon M. DeHayes and David Stark, Eds., American Society for Testing and Materials, Philadelphia, 1994.

ABSTRACT: The capillary porosity of hardened concrete can be revealed by impregnation of the sample by a fluorescent epoxy resin.
The paper presents a method of measurement of the cement paste fluorescence (i.e. the W/C ratio) by image analysis coupled with an automatical statistical interpretation of the data.
Measurements were performed on thin sections of concrete of known W/C ratio with ASTM type I and ASTM type I-S cement.
The tests allow to conclude for a significant influence of the cement type on the relation between fluorescence and W/C ratio.
Further it is shown that the estimation of the W/C ratio depends on the degree of hydration of the concrete.
It is concluded that the measurement of the fluorescence gives an estimate of the capillary porosity rather than an estimate of the real W/C ratio.

KEYWORDS: Water/cement ratio, fluorescence, thin section, image analysis, cement type, degree of hydration, analysis of variance.

INTRODUCTION

The knowledge of the water/cement ratio is very useful in the assessment of the quality of hardened concrete.

A microscopical method of estimation of this parameter has been developed in Denmark [1,2,3]. The estimation is performed on thin sections impregnated under vacuum with a fluorescent epoxy resin that fills the capillary pores and makes the cement paste shine when examined with blue-violet light.The intensity of the cement paste fluorescence rises with the capillary porosity and hence with the W/C ratio. A comparison of a thin section from an unknown sample with thin sections of known W/C ratio's concrete samples,allows an evaluation of the unknown W/C ratio.

A trial of quantitative evaluation of the W/C ratio by fluorescence microscopy has been made by Mayfield [4] using a photo-diode integral amplifier for the measurement of light.
On the other hand,Wirgot and Van Cauwelaert have developed a method of measurement of the cement paste fluorescence by image analysis coupled with the interpretation of the data by use of the beta statistic [5].

After a brief recall of this method principle,measurements of hardened concrete cement paste fluorescence are presented and their significance is discussed.

---

[1] Belgian Cement Industry Collective Research Centre, Brussels, Belgium

TEST PROGRAM

In order to evaluate the effects of the water/cement ratio, of the cement type and of the degree of hydration on the capillary porosity as it is revealed by the cement paste fluorescence, 3 series of concrete mixes were prepared: a first series (series A) with portland cement and water/cement ratio between 0.35 and 0.70, a second series (series B) with slag cement and water/cement ratio between 0.40 and 0.70 and a third series (series C) with portland cement, water/cement ratio between 0.40 and 0.60 and 3 curing conditions (to produce different degrees of hydration).

MATERIALS, MIX CHARACTERISTICS AND OPERATING PROCEDURES

The Series A concrete mixes were made with ordinary portland cement (ASTM Type I-Belgian type P-40), rounded river gravel (4-28 mm) as coarse aggregate and Rhenan sand (0.18-5 mm) as fine aggregate. The aggregate gradings are shown in Table 1 (coarse aggregate) and Table 2 (fine aggregate). Superplasticising-high range water-reducing admixture was used for the A-1 mix in the proportion of 4 kg admixture for 400 kg cement.

TABLE 1 - Coarse aggregate grading.

Sieve size	Cumulative percentage retained
28.0 mm	1.4
16.0 mm	42.2
10.0 mm	73.0
6.3 mm	89.0
4.0 mm	97.0
2.0 mm	98.5

TABLE 2 - Fine aggregate grading

Sieve Size	Cumulative percentage retained
4.0 mm	3.1
2.0 mm	8.2
1.0 mm	17.8
0.63 mm	33.0
0.25 mm	89.4

Mix characteristics and properties of fresh concrete are given in Table 3. Air content and consistency were measured respectively according to the Belgian standards NBN B15-208 and NBN B15-233. Two 158 mm edge cubes and two 100- by 100 - by 400-mm prisms were cast from each of the mixes.After 24 hours of air curing, the samples were demolded and immersed in water at 20°C for a period of 56 days. Afterwards, they were conserved at 20°C and 60 % RH.

TABLE 3 - Mix characteristics and properties of fresh concrete(Series A).

Mix	Mix characteristics					Properties of fresh concrete		
	Coarse aggregate content kg/m³	Sand content kg/m³	Cement content kg/m³	Water content kg/m³	W/C ratio	Air content %	Unit Mass kg/m³	Consis-tency (Flow-test)
A-1	1300	560	400	140	0.35	1.5	2420	1.60
A-2	1270	560	400	160	0.40	2.4	2375	1.15
A-3	1240	590	380	171	0.45	1.2	2387	1.73
A-4	1225	615	360	180	0.50	1.0	2380	1.98
A-5	1220	650	330	181.5	0.55	1.4	2365	1.98
A-6	1200	665	310	186	0.60	1.0	2360	2.22
A-7	1190	700	280	182	0.65	0.9	2365	2.29
A-8	1185	730	260	182	0.70	1.7	2340	2.11

The Series B concrete mixes were prepared with a slag cement (ASTM Type I-S -Belgian Type HK-40) and the same aggregates as those used in the Series A mixes. Mix characteristics and properties of fresh concrete are given in Table 4. Consistency was measured according to the Belgian standard NBN B15-233. Two 158 mm edge cubes and two 100- by 100- by 400-mm prisms were cast from each of the mixes. After 24 hours of air curing, the samples were demolded and immersed in water at 20°C for a period of minimum 56 days. Afterwards, they were cured at 20°C and 60 % RH.

TABLE 4 - Mix characteristics and properties of fresh concrete(Series B).

Mix	Mix characteristics					Properties of fresh concrete	
	Coarse aggregate content kg/m³	Sand content kg/m³	Cement content kg/m³	Water content kg/m³	W/C ratio	Unit Mass kg/m³	Consis-tency (Flow-test)
B-1	1240	560	420	168	0.40	2385	1.53
B-2	1220	600	390	175.5	0.45	2375	1.77
B-3	1210	630	360	180	0.50	2365	2.00
B-4	1160	670	340	187	0.55	2355	2.17
B-5	1140	690	320	192	0.60	2350	2.45
B-6	1160	710	290	188.5	0.65	2310	2.48
B-7	1160	740	260	182	0.70	2335	2.30

The series C concrete mixes were prepared with a portland cement similar to the series A cement and with the same aggregates than the series A mixes. Mix characteristics and properties of fresh concrete are shown in Table 5. Air content and consistency were measured following the Belgian standards NBN B15-208 and NBN B15-233. Three 158 mm edge cubes and three 100- by 100- by 400-mm prisms were cast from each of the 3 mixes, air cured for 24 hours and demolded. The samples were then cured by 3 different modes: - either air curing at 20°C, 60% RH (mode 1) - or 14 days water curing at 20°C followed by air curing at 20°C and minimum 95% RH (mode 2)- or 56 days water curing at 20°C followed by air curing at 20°C and 95 % RH (mode 3).

TABLE 5 - Mix characteristics and properties of fresh concrete(Series C).

Mix	Mix characteristics					Properties of fresh concrete	
	Coarse aggregate content kg/m³	Sand content kg/m³	Cement content kg/m³	Water content kg/m³	W/C ratio	Air content %	Unit Mass kg/m³
C-1	1280	545	400	160	0.40	1.0	2400
C-2	1220	615	360	180	0.50	1.1	2375
C-3	1110	725	320	192	0.60	1.3	2350

THIN SECTIONS PREPARATION AND EXAMINATION

The 30- by 45-mm thin sections are prepared by following a scrupulously controlled process: the concrete specimens are oven dried at 105°C for 3 days in an inert atmosphere, impregnated under monitored vacuum by a fluorescent epoxy resin and finally lapped to a 25 microns thickness and protected by a cover-glass.

The thin sections are examined with a petrographic microscope in polarized light and reflected fluorescence modes. The reflected light is furnished by a 100 W halogen source fed with direct current. The excitation radiation is comprised between 420 and 490 nm (violet-blue light). A barrier filter stops radiations above 530 nm.
A CCD black and white camera sends the image of the microscope field to a Tracor TN 8502-C image analysis system. The statistical analysis (approximation by a beta distribution) of the data issued from the Tracor computer is performed on a personal computer.

FLUORESCENCE MEASUREMENT PROCEDURE

The blue-violet light causes a fluorescence of the air voids and of the cement paste.The black and white image of the microscope field (Fig.1) comprises black(non fluorescent) aggregate grains,grey fluorescent cement paste,and white fluorescent air voids.

Figure 1
Black and white image of fluorescent thin section
The cement paste appears in grey while
the air voids are white and the aggregate grains are black

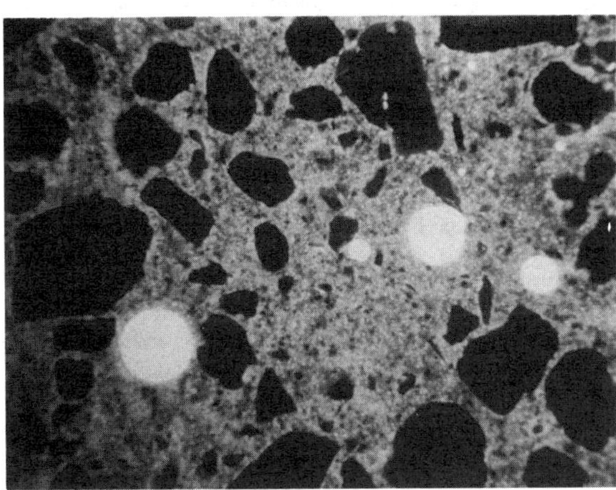

1 mm

Once digitalized, the image is composed of 512 by 512 pixels, each pixel having a grey level value comprised between 0 and 255.
The grey level histogram of that kind of image shows 3 distributions (Fig.2) corresponding respectively to the sand grains (A distribution), the cement paste (B distribution) and the air voids (C distribution).

Figure 2
Grey level histogram of Figure 1

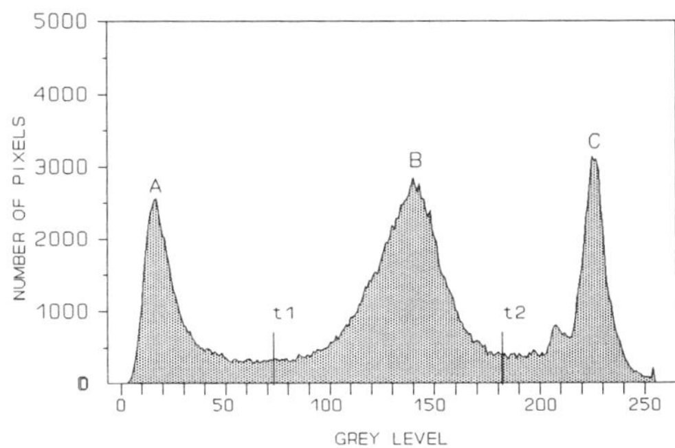

The method consists in associating a characteristic grey level to the distribution corresponding to the cement paste (B distribution of Fig.2).
A binary image is created by selecting the cement paste area of the black and white image. The frequency histogram of the B distribution, comprised between two threshold values t1 and t2 (Fig.2), is so isolated and sent to a personal computer where it is approximated with an appropriate beta-distribution (Type I of Pearson's system) whose mean value is supposed to represent the characteristic grey level of the cement paste. The approximation by the beta distribution also allows the determination of the statistical parameters of the B distribution [5]. It is assumed that the mean of the B distribution is a measurement of the cement paste capillary porosity and hence of the concrete water/cement ratio:the greater the w/c ratio,the brighter the cement paste fluorescence,and the higher the characteristic grey level.

Twelve microscopic fields of 2 mm^2 individual area are measured within a thin section. The mean of the twelve characteristic grey level values is supposed to represent the W/C ratio of the thin section.

The repeatability of the method is very good: the coefficient of variation of the measurement of the mean characteristic grey level value of a thin section is about 0.8 %.

RESULTS

Series A

After two years, two thin sections were prepared within a cube of each mix. The mean characteristic grey level of the cement paste resulting from the measure of 12 fields within each thin section is reported versus the W/C ratio in Figure 3. The mean values are shown in Table 6 and, for the interested reader, the individual values are shown in Appendix 1.

Figure 3
Grey level vs W/C ratio for Series A concrete

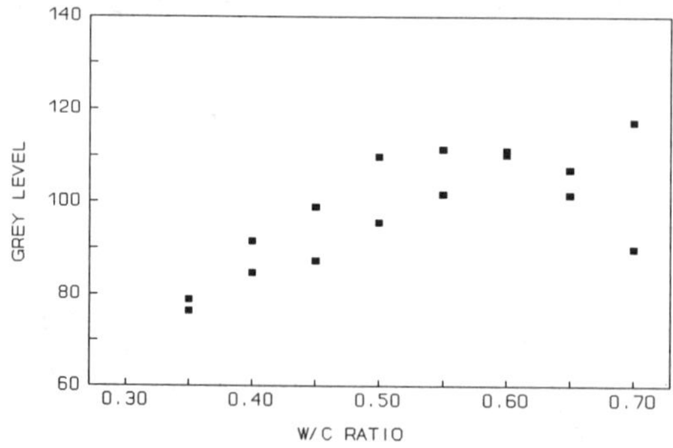

TABLE 6 - Characteristic grey level values of the Series A thin sections

Mix	Thin section	W/C ratio	Characteristic Grey level value	
			Mean	Coefficient of variation (%)
A-1	A-1/1	0.35	76.37	5.55
A-1	A-1/2	0.35	78.82	3.39
A-2	A-2/1	0.40	84.69	5.50
A-2	A-2/2	0.40	91.53	9.65
A-3	A-3/1	0.45	87.30	7.15
A-3	A-3/2	0.45	98.95	6.86
A-4	A-4/1	0.50	109.9	5.82
A-4	A-4/2	0.50	95.58	5.13
A-5	A-5/1	0.55	111.52	4.40
A-5	A-5/2	0.55	101.79	7.08
A-6	A-6/1	0.60	111.17	6.28
A-6	A-6/2	0.60	110.25	5.67
A-7	A-7/1	0.65	101.67	7.87
A-7	A-7/2	0.65	107.12	3.70
A-8	A-8/1	0.70	89.92	8.75
A-8	A-8/2	0.70	117.24	7.31

As the variability of the fluorescence is quite high within and between the thin sections (Table 6),it was decided to test the statistical representativity of the mean characteristic grey level of one thin section.
5 thin sections were extracted side by side at the centre of a slab sawn at mid-height of a concrete cube coming from the A-4 mix.
The mean grey level values resulting from the measure of these thin sections are shown in Table 7 while the individual values are given in Table 13.

TABLE 7 - Characteristic grey level values of the Series A thin sections

Mix	Thin section	W/C ratio	Characteristic Grey level value	
			Mean	Coefficient of variation (%)
A-4	A-4/3	0.50	108.99	4.89
A-4	A-4/4	0.50	114.66	4.47
A-4	A-4/5	0.50	117.32	4.12
A-4	A-4/6	0.50	119.65	2.47
A-4	A-4/7	0.50	124.38	4.90

Series B

After one year, two thin sections were prepared within a cube of each mix. The mean characteristic grey level of the cement paste resulting from the measure of 12 fields within each thin section is reported versus the W/C ratio in Figure 4. The mean values are shown in Table 8 and, for the interested reader, the individual values are shown in Appendix 2.

Figure 4
Grey level vs W/C ratio for Series B concrete

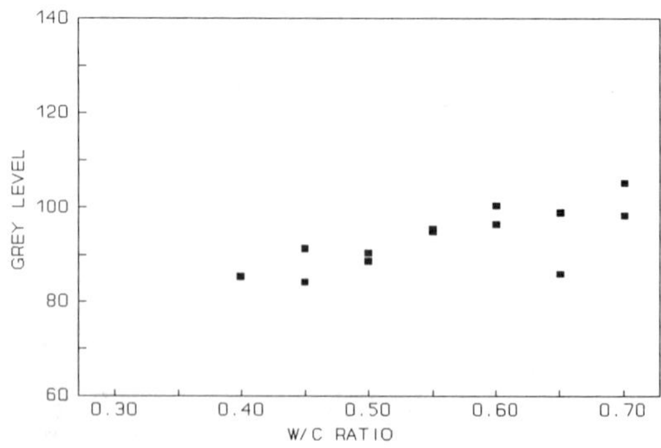

TABLE 8 - Characteristic grey level values of the Series B thin sections

Mix	Thin section	W/C ratio	Characteristic Grey level value	
			Mean	Coefficient of variation (%)
B-1	B-1/1	0.40	85.73	3.77
B-2	B-2/1	0.45	84.53	5.68
B-2	B-2/2	0.45	91.75	2.66
B-3	B-3/1	0.50	89.09	3.30
B-3	B-3/2	0.50	90.83	4.29
B-4	B-4/1	0.55	95.85	3.36
B-4	B-4/2	0.55	95.41	4.76
B-5	B-5/1	0.60	100.78	6.25
B-5	B-5/2	0.60	96.83	3.08
B-6	B-6/1	0.65	99.29	4.46
B-6	B-6/2	0.65	86.38	5.06
B-7	B-7/1	0.70	105.51	6.01
B-7	B-7/2	0.70	98.67	3.49

Series C

After one year, two thin sections were prepared within a cube of each mix. The results of the measures are shown in Table 9.
The fluorescence of some thin sections is so weak or so heterogeneous that the measurement is impossible.

TABLE 9 - Characteristic grey level values of the Series C thin sections

Mix	Curing mode	Thin section	W/C ratio	Characteristic Grey level value	
				Mean	Coefficient of variation (%)
C-1	1	C-1/1a	0.40	87.53	3.67
C-1	1	C-1/1b	0.40	84.61	2.77
C-1	2	C-1/2a	0.40	-	-
C-1	2	C-1/2b	0.40	-	-
C-1	3	C-1/3a	0.40	-	-
C-1	3	C-1/3b	0.40	-	-
C-2	1	C-2/1a	0.50	94.68	3.01
C-2	1	C-2/1b	0.50	97.93	3.17
C-2	2	C-2/2a	0.50	-	-
C-2	2	C-2/2b	0.50	-	-
C-2	3	C-2/3a	0.50	-	-
C-2	3	C-2/3b	0.50	-	-
C-3	1	C-3/1a	0.60	107.40	3.27
C-3	1	C-3/1b	0.60	105.02	1.76
C-3	2	C-3/2a	0.60	96.89	2.98
C-3	2	C-3/2b	0.60	102.06	2.02
C-3	3	C-3/3a	0.60	96.25	3.78
C-3	3	C-3/3b	0.60	101.63	3.82

- : Fluorescence too weak to be measurable

DISCUSSION

Series A and B

The results of the measurements of the Series A and B thin sections are given in Table 10.

TABLE 10 - Grey levels vs W/C ratio's

W/C	0.35	0.40	0.45	0,50	0,55	0,60	0,65	0,70
Series A								
Grey level								
series 1	76.37	84.69	87.30	109.90	111.52	111.17	101.67	89.92
series 2	78.82	91.53	98.95	95.58	101.79	110.25	107.12	117.24
Series B								
Grey level								
series 1	-	85.73	84.53	89.09	95.85	100.78	99.29	105.51
series 2	-	85.73	91.75	90.83	95.41	96.83	86.38	98.67

In order to compare the results between Series 1 and Series 2 of each concrete we have performed a two variables variance analysis, on the data of Series A and Series B respectively.
The results of the analysis are given in Table 11 for Series A concrete and in Table 12 for Series B concrete.

TABLE 11 - Variance analysis for Series A concrete

Source of variability	Sum of squares	Degrees of freedom	Mean square	Test	$F_{95\%}$
Between columns	1736	7	248	248/83=3.0	3.79
Between series	52	1	52	52/83=0.6	5.59
Residual	581	7	83		
Total	2369	15	158		

TABLE 12 - Variance analysis for Series B concrete

Source of variability	Sum of squares	Degrees of freedom	Mean square	Test	$F_{95\%}$
Between columns	417	6	70	70/21=3.3	4.28
Between series	17	1	17	17/21=0.8	5.99
Residual	126	6	21		
Total	560	13	43		

It can be concluded from the analysis that for Series A and Series B concrete there is no significative difference between the results of the two sets (Series 1 and Series 2) of thin sections: the test values 0.6 and 0.8 are very low against the significant F-values (5.59 and 5.99). Therefore one could be inclined to conclude that there exists a good relation between the grey-level and the W/C ratio, independant from the examined sample. However when considering the variability between columns (i.e. between the W/C ratio's), we must surprisingly state that here also there is no significative difference between the results, which practically means that apparently there is no relation between the analyzed variables.

In order to obtain a better insight in this quite unexpected phenomenon we have compared the results obtained on five thin sections taken out of a concrete sample coming from the A-4 mix, with a W/C ratio of 0.50.
The grey levels were measured on 12 fields for each thin section.
The results of the measurements are given in Table 13.

TABLE 13 - Grey levels of 5 adjoining thin sections.

Field	Thin Section 1	Thin Section 2	Thin Section 3	Thin Section 4	Thin Section 5
1	117.13	118.82	104.14	116.83	111.62
2	110.37	128.37	119.24	121.47	113.46
3	111.19	113.20	118.17	119.25	113.35
4	117.82	133.62	111.94	116.02	116.71
5	115.42	133.55	112.00	116.27	110.91
6	126.58	124.80	114.73	119.66	111.63
7	121.38	124.34	111.11	115.87	112.63
8	119.68	120.27	113.01	123.60	107.26
9	120.95	124.75	120.08	123.14	99.18
10	112.53	120.93	123.60	123.69	103.67
11	117.50	125.55	112.39	120.84	103.19
12	117.32	124.38	115.56	119.10	104.29

In order to compare the results between the 5 thin sections we have performed a one variable (grey levels of each thin section) variance analysis. The results are presented in Table 14.

TABLE 14 - Variance analysis for 5 adjoining thin sections

Source of variability	Sum of squares	Degrees of freedom	Mean square	Test	$F_{99\%}$
Between columns	1519	4	380	380/25=15.2	3.70
Intra-columns	1303	53	25		
Total	2822	57	50		

The test value of 15.2 is much higher than the F-value by a probability of 99 %. Hence the results obtained on the 5 thin sections differ significantly with a very high degree of probability and thus the null hypothesis test (are the 5 mean values equal ?) must be rejected.

Therefore it can be concluded that the grey level of a thin section is only representative of the W/C ratio of that peculiar thin section. Thus if we want to compare different samples of concrete, we must analyze more than one thin section per sample.

In order to test the influence of the cement type on the capillary porosity revealed by the cement paste fluorescence, we perform now a two variable variance analysis on the integrality of the results of Table 10, in order to compare Series A concrete with Series B concrete.
The results of the analysis are given in Table 15.

TABLE 15 - Variance analysis for series A concrete vs series B concrete

Source of variability	Sum of squares	Degrees of freedom	Mean square	Test	$F_{97,5\%}$
Between columns	1023	6	171	171/55=3.1	3.50
Between series	450	1	450	450/55=8.2	6.30
Interaction	143	6	24	24/55=0.4	3.50
Intra-combination	772	14	55		
Total	2388	27	88		

Again it can be concluded that the W/C ratio has no significative influence on the grey level. However the test between series (i.e. between the two different concretes) is now significative at the 95 % level. Hence there seems to be a significative influence of the type of cement on the cement paste characteristic grey level of a concrete thin section.

The lack of influence of the W/C ratio on the grey level is very surprising. As bleeding has occured within the 0.65 and 0.70 mixes, we suppose that the real W/C ratio's of the concrete samples are not the computed W/C ratio's.
Therefore, a new analysis of variance has been achieved without the W/C = 0.65 and 0.70 grey level values.
The results are shown in Tables 16, 17 and 18 corresponding respectively to the Tables 11, 12 and 15.

TABLE 16
Variance analysis for Series A concrete without W/C = 0.65 and 0.70 grey level values

Source of variability	Sum of squares	Degrees of freedom	Mean square	Test	$F_{95\%}$
Between columns	1566	5	313	313/49=6.39	5.05
Between series	1,35	1	1,35	1/49=0.02	6.61
Residual	243	5	49		
Total	1810	11	165		

TABLE 17
Variance analysis for Series B concrete without W/C = 0.65 and 0.70 grey level values

Source of variability	Sum of squares	Degrees of freedom	Mean square	Test	$F_{95\%}$
Between columns	234	4	59	59/8=7.38	6.39
Between series	2	1	2	2/8=0.25	7.71
Residual	33	4	8		
Total	269	9	30		

TABLE 18
Variance analysis for Series A concrete vs series B concrete without W/C = 0.65 and 0.70 grey level values

Source of variability	Sum of squares	Degrees of freedom	Mean square	Test	$F_{95\%}$
Between columns	858	4	215	215/28=7.68	4.47
Between series	371	1	371	371/28=13.25	6.94
Interaction	86	4	22	22/28=0.78	4.47
Intra-combination	277	10	28		
Total	1592	19	84		

This time, the results of the analysis of variance show a significative influence of the W/C ratio on the characteristic grey level of the cement paste.

The measurements also show that, for a given W/C ratio, the grey level values of the Series B thin sections are lower than the Series A thin sections for 85 % cases.
One can interpretate this as the result either of a finer pore structure or of a lower capillary porosity of the slag cement paste.
This particularity of the slag cement pastes has already been mentioned by authors [6,7,8] using other analytical techniques.

Series C

The measurements of the Series C thin sections show a systematic decrease of the cement paste fluorescence with an increasing degree of hydration (Table 9).
Indeed, the degree of hydration, revealed by the amount of unhydrated clinker grains, increases with the relative humidity of the curing mode.

Figures 5 and 6 show the fluorescences of the C-2/1a (W/C = 0.50,curing mode 1) and C-2/3a (W/C = 0.50,curing mode 3) thin sections : the increased production of C-S-H in curing mode 3 reduces the capillary porosity of the cement paste and thus the fluorescence of the thin section.

Figure 5
Black and white image of
fluorescent thin section N° C-2/1a
W/C = O.50 - Curing mode 1

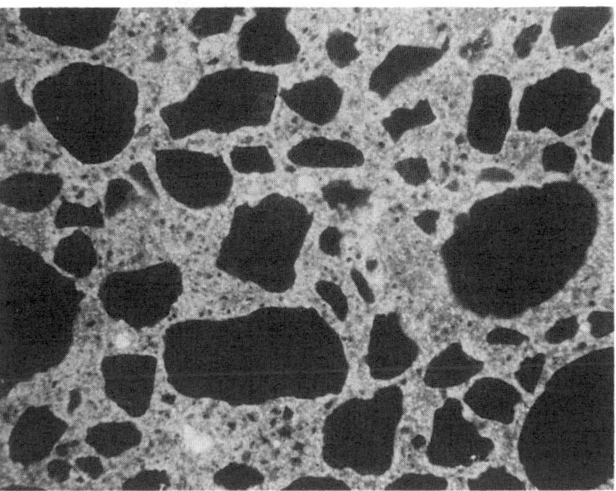

1 mm

Figure 6
Black and white image of
fluorescent thin section N° C-2/3a
W/C = 0.50 - Curing mode 3

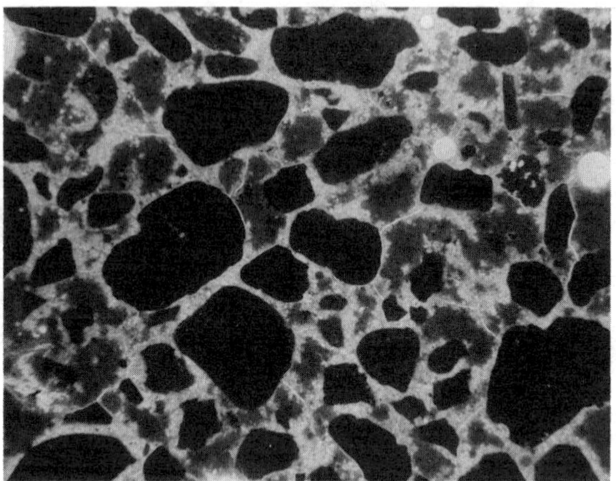

1 mm

CONCLUSION

The described method of measurement of the cement paste fluorescence can be used to assess the fluorescence and, hence, the capillary porosity of a concrete thin section. Indeed, the repeatability of the measurement is excellent.

The effects of the W/C ratio, of the cement type and the degree of hydration on the capillary porosity are evaluated by measuring the fluorescence of thin sections extracted from 3 series of concrete mixtures.

Measurements performed on thin sections extracted from portland cement and slag cement concretes are first interpreted using two variables (W/C versus grey level) variance analysis. The conclusion is that, for both series of concretes, the W/C ratio has no influence on the mean grey level of the thin section.
This unexpected phenomenon can be explained by the influence of the results of the thin sections extracted from the W/C = 0.65 and 0.70 mixes because of a probable inaccurate value of W due to bleeding.

Then, ignoring the values of W/C = 0.65 and 0.70, it can be concluded from a second variance analysis that, indeed, the W/C ratio has a significative influence on the mean grey level of the thin section.

This analysis also demonstrates that the cement type has an influence on the mean grey level.

Further, one variable (grey level for W/C = 0.50) variance analysis shows that the variation between the results obtained on adjoining thin sections of a same concrete sample is too high to allow a sufficiently accurate correlation between the mean grey level of a thin section and the W/C ratio of the concrete sample.

Finally, measurements performed on thin sections extracted from concretes cured by 3 different modes show the influence of the degree of hydration on the grey level value of the cement paste.

Hence it can be concluded that, although the cement paste fluorescence of concrete thin sections can significantly be related to the W/C ratio of the concrete, it also depends on the used cement and on the degree of hydration of the cement.

It also should be noticed that, due to the scatter of the fluorescence measurements within and between thin sections, more than one thin section per sample is required to allow a comparison of concrete samples.

ACKNOWLEDGEMENTS

The authors would like to acknowledge M. De Lanève and J.M. Siscot for their assistance during the research.
The financial support of the "Institut pour l'Encouragement de la Recherche dans l'Industrie et l'Agriculture" is also granted.

Appendix 1 - Individual grey level values (Series A)

A-1/1	A-1/2	A-2/1	A-2/2	A-3/1	A-3/2	A-4/1	A-4/2
72.52	78.65	81.26	101.73	82.70	103.08	106.25	92.81
69.50	82.57	90.68	103.66	88.00	100.72	106.64	91.29
73.45	82.07	86.50	100.88	83.15	101.51	111.23	88.36
71.68	80.40	89.19	95.53	87.11	104.08	109.93	89.55
77.67	81.48	86.84	78.41	87.39	103.78	106.49	97.66
79.02	80.54	87.71	81.67	95.17	105.36	108.91	100.43
81.03	77.62	87.38	78.69	88.54	93.75	101.99	101.11
80.77	79.58	86.57	97.82	79.20	96.26	102.70	101.46
80.16	75.44	84.19	92.78	82.32	96.46	116.08	100.43
78.10	76.06	83.33	89.20	95.78	81.09	122.00	95.80
80.66	75.16	75.53	86.89	98.00	97.12	120.01	92.52
71.89	76.26	77.13	91.06	80.29	104.21	106.60	-

A-5/1	A-5/2	A-6/1	A-6/2	A-7/1	A-7/2	A-8/1	A-8/2
113.37	100.37	104.94	108.88	90.91	101.87	92.45	114.67
110.76	108.66	105.82	108.78	99.41	106.37	91.49	110.83
121.24	110.07	103.08	110.60	96.23	105.71	85.94	122.15
111.83	102.96	105.59	118.67	100.51	113.33	88.36	134.35
102.81	112.68	108.44	114.31	106.31	100.97	90.36	113.85
112.08	91.48	105.27	118.17	113.83	108.71	102.31	113.79
109.01	97.30	120.09	113.44	113.87	103.72	95.77	131.73
113.99	91.03	124.25	114.79	107.49	105.96	101.34	109.87
116.99	99.84	113.19	109.18	94.87	106.97	89.34	115.24
107.24	95.74	119.72	107.39	90.15	111.80	84.07	108.09
112.69	108.89	111.27	99.08	100.34	112.48	73.60	110.39
106.21	102.42	112.41	99.75	106.12	107.52	84.05	121.95

Appendix 2 - Individual grey level values (Series B)

B-1/1	B-2/1	B-2/2	B-3/1	B-3/2	B-4/1	B-4/2
86.00	89.88	92.20	89.17	83.89	91.86	89.59
88.01	91.11	95.86	92.34	88.69	88.52	98.95
87.85	89.29	91.70	88.61	92.61	98.69	98.75
83.03	82.44	91.81	90.34	97.06	99.36	99.91
82.23	84.54	94.68	91.03	92.46	97.63	100.35
81.81	83.76	92.52	91.24	88.99	96.16	90.10
85.04	84.34	93.75	84.89	94.07	93.84	86.51
83.89	87.91	89.67	91.34	90.74	94.90	97.20
83.20	79.43	89.64	90.91	96.61	95.30	98.90
92.64	74.22	89.68	84.19	88.98	99.08	95.10
89.22	81.63	87.69	84.38	86.80	97.96	95.41
85.80	85.86	-	90.64	89.11	96.84	94.09

B-5/1	B-5/2	B-6/1	B-6/2	B-7/1	B-7/2
101.21	89.28	97.13	87.08	111.88	94.41
106.16	95.23	106.35	93.32	96.43	100.10
105.08	96.81	94.74	91.56	97.88	93.40
110.64	98.48	95.10	83.19	103.67	99.18
101.73	97.07	98.29	88.86	102.75	99.78
91.19	99.27	104.55	90.07	102.02	99.89
94.70	96.29	97.72	88.30	110.25	99.36
96.49	99.08	101.23	82.56	109.26	103.95
92.32	99.40	96.32	82.69	119.14	104.41
98.41	99.46	106.05	84.73	106.77	96.58
102.64	97.81	100.44	78.01	101.98	95.00
108.75	93.74	93.59	86.18	104.09	98.03

REFERENCES

[1] Christensen,P., Gudmundsson,H., Thaulow,N., Damgard Jensen, A., Chatterji,S., "Struktur- og bestandelsanalyse af beton", Nordisk Betong, 3-1979, pp.4-10.

[2] Jensen,A.D., "Investigation of concrete by analysis of thin sections", IABSE Symposium on strengthening of building structures, Venice 1983, pp.109-115.

[3] Thaulow,N., Damgard Jensen,A., Chatterji, S., Christensen,P., Gudmundsson,H., "Estimation of the compressive strength of concrete samples by means of fluorescence microscopy", Nordisk Betong, 2-4-1982, pp.51-52.

[4] Mayfield,B., "The quantitative evaluation of the water/cement ratio using fluorescence microscopy", Magazine of Concrete Research, Vol.42, No.150, pp.45-49, 1990.

[5] Wirgot,S., Van Cauwelaert,F., "The measurement of impregnated cement paste fluorescence by means of image analysis", Proceedings of the Third Euroseminar on microscopy applied to building materials, Barcelona, September 1991.

[6] Roy,D.M., Idorn,G.M., "Hydration, Structure and Properties of Blast Furnace Slag Cements, Mortars, and Concrete", Journal of the American Concrete Institute, Vol.79, No.6, pp.444-457, 1982.

[7] Manmohan,D., Mehta,P.K., "Influence of Pozzolanic, Slag, and Chemical Admixtures on Pore Size Distribution and Permeability of Hardened Cement Pastes", Cement, Concrete, and Aggregates, Vol.3, No.1, pp.63-67, 1981.

[8] Marion,A.M., "Contribution à l'étude de l'influence des ajouts minéraux sur la microstructure du ciment hydraté", Thèse de Doctorat en Sciences, Université Libre de Bruxelles, 1989.

John Cahill,[1] Jill C. Dolan,[2] and Peter W. Inward[1]

THE IDENTIFICATION AND MEASUREMENT OF ENTRAINED AIR IN CONCRETE USING IMAGE ANALYSIS

REFERENCE: Cahill, J., Dolan, J. C., and Inward, P. W., **"The Identification and Measurement of Entrained Air in Concrete Using Image Analysis,"** Petrography of Cementitious Materials, ASTM STP 1215, Sharon M. DeHayes and David Stark, Eds., American Society for Testing and Materials, Philadelphia, 1994.

ABSTRACT: The use of ASTM C457 ("linear traverse" or "modified point-count methods") for microscopical determination of air-void content and parameters of the air-void system, in hardened concrete has become widespread in the prediction of the durability of concrete in freeze-thaw conditions, and in assessing concrete that has failed in service. More recently a new European Standard, based on linear traverse measurements has been adopted as an efficacy test for air entraining agents. The linear traverse method described in both standards involves identification, counting and, in the Euro Standard, measurement of air-voids. It is laborious, tedious and subjective and the error of the method has not been rigorously assessed. We have explored the possibility of using Image Analysis in conjunction with the linear traverse method to develop less laborious and subjective measurement methods. A number of methods of visualisation of air-voids have been applied to polished samples of hardened concrete which had previously been assessed by the standard linear traverse method. The quality of the surface preparation required for Image Analysis was found to be even more critical than for the standard manual method.

KEYWORDS: Spacing factor, air entrainment, concrete, Image Analysis

The ability of concrete containing between 4% and 6% by volume of entrained air to resist damage from freeze thaw cycling is well documented. Various theories have been advanced regarding the physical mechanism by which the concrete is protected. Conceptually simple is the

[1]Scientist, Unilever Research, Port Sunlight Laboratory, Bebington, Wirral, UK.

[2]Research and Development Manager, Cormix Construction Chemicals, Warrington, UK.

idea of air-voids acting as expansion chambers, minimising the damage caused by pressure generated from the expansion of pore water on freezing. This is consistent with current specifications relating freeze thaw durability to air-void spacing factor, a parameter relating the maximum distance at any one point in the hardened cement paste from the periphery of an air-void.

The American Standard ASTM C457-90 [1] describes a modified point-count and linear traverse method for measurement of the spacing factor. More recently a new European Standard ENV 480 Part II [2] has introduced the mandatory measurement of spacing factor, superseding in the UK, the previous British Standard based on the measurement of total air content of fresh concrete. This new European Standard is similar to the ASTM Linear Traverse method. In both methods, polished sections of concrete are examined microscopically along a series of regularly spaced lines of traverse. The essential feature of entrained air-voids, which allows their identification, is that they are spherical and, therefore, circular in section, whereas other features will generally be irregular in shape. Every air-void encountered on the traverse must be recognised and recorded and in the case of the linear traverse method, its chord length measured. It will be readily appreciated that these methods are labour intensive and both tedious and tiring for the operator. There is, therefore, a clear need for an automated procedure which would eliminate most of the tedium and some of the subjectivity involved in current procedures.

This study is an attempt to use modern Image Analysis techniques in the development of an automated process designed to reproduce the analysis according to the linear traverse method. Results obtained by Image Analysis are compared with those obtained by experienced operators using the standard linear traverse method, for a series of Cormix air entraining agents.

LINEAR TRAVERSE METHOD (SUMMARY)

Cubes of hardened concrete are sectioned perpendicular to the original free surface to produce specimens for analysis. These specimens are then ground to produce a smooth flat surface finish suitable for microscopical investigation. The polished specimen is mounted under a microscope with a magnification of 100 +/- 10x. The air-void structure is examined by scanning along a series of traverse-lines running parallel to the original free surface. The number of air-voids intersected by the traverse-lines are recorded, as are the individual chord lengths of the traverse across the air-voids. If, in spite of careful grinding, the edges of the voids are broken and such a breakage lies on a traverse line then the completed circular section shall be used as the basis for determining the chord length. A mathematical analysis of the recorded data then allows a description of the air-void system in terms of the following parameters:-

Total air content. The proportion of the total volume of concrete taken up by air-voids, expressed as a percentage by volume.

Specific surface of entrained air. A calculated parameter representing the total surface area of air-voids divided by their volume.

Spacing factor. A calculated parameter related to the maximum distance in the cement paste from the circumference of an air-void.

Air-void size distribution. A set of calculated values of the number of air-voids of various diameters within the cement paste. Calculated from the distribution of chord lengths during the traverse procedure.

Micro air content. A calculated parameter representing the air content attributed to air-voids of 0.3 mm diameter or less.

THE IMAGE ANALYSIS METHOD

The image acquisition and analysis system is shown schematically in Fig. 1. It comprises a motorised X, Y stage, a microscope with focus control, a high resolution monochrome camera, a motor control box, and the Kontron IBAS Image Analysis system.

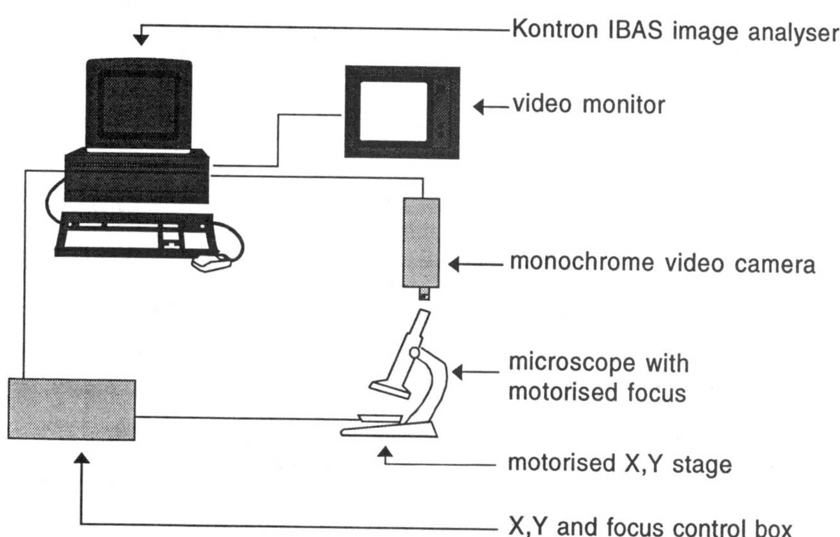

Fig. 1 Schematic of image aquisition and analysis system

The traverse of the camera over the polished surface of the concrete mimics that defined in the conventional linear traverse methods. Each of the images produced by the camera is analysed by placing it on a fine [512 x 512] grid. The system examines every square on this grid and allocates to each square, a number representing its place on a numerical scale ranging from 0 [black] to 255 [pure white] via intermediate numbers representing shades of grey. Once the image has been digitised in this way, the information is stored and can be analysed mathematically for example shapes or points of dark or light and the distance between them. The scope for analysis is limited only by the capabilities of available computer software.

Identification of Voids

Three methods were considered for identification of air-voids by accentuating the contrast between air-voids and the concrete surface.

1. Low angle illumination to shadow voids. This method is analogous to that used in the manual technique. Both circularity, shadow and shape factors are retained within the image.
2. This method is suggested in the European Standard. [2] The surface is first carefully inked black to reduce contrast between aggregate and paste and allowed to dry. The voids are then highlighted by working into them either a white zinc paste or alumina powder. In perfect circumstances it would produce white circles on a black background, ideal for imaging.
3. Impregnation with epoxy resin of voids in the surface of concrete, followed by flatting back to the original surface. This method is more laborious but if a colour-camera is available, it allows a choice of colour in the resin to contrast with the paste and aggregate. In addition to highlighting the voids the resin provides support to minimise damage to the edges of the voids during surface-preparation.

EXPERIMENTAL SECTION

Surface Preparation

As in the conventional method, a highly polished concrete surface was required in order to offer the best opportunity for discrimination between air-voids and other surface features. The methods used were essentially as described in the published ASTM [1] and European [2] Standards. Surfaces cut with a diamond saw were polished with successively finer grades of silicon carbide abrasive.

Acquisition and Measurement of Void Images

For the automatic method the image is captured, digitised and analysed for contrast. The microscope was set to a magnification such that an image of 512 x 512 picture elements covered an area of 500 x 500 micrometers of the sample. The focus, under computer control is adjusted until maximum contrast and hence best focus is achieved. The image is then acquired and any void within the image is identified solely on the criterion of intensity of grey value where 0 = black.. 255 = white.

All occurrences of grey value above a threshold of eg. 220 are assumed part of a void or voids and are turned white. The background, values 220 or below, are converted to black, so producing a binary black/white image. A traverse line is then placed across the centre of the binary image and overlap of the traverse line with any air-void yields chord intercepts. These chords are measured and tagged on the criteria of touching the surrounding frame, left side, right side, both sides or neither side. The sample is then moved exactly one field to the side and automatically re-focussed, identified and void chords measured. Once a complete traverse of the sample is finished the data set may be analysed.

To obtain the true length of the chords that have overflowed one or more fields of view, it is necessary for the data to be rebuilt. This is accomplished by using the various tags to join overflows. Finally the data are imported into a spreadsheet used to calculate the parameters defined earlier.

Calculations

The method of calculation of air parameters is taken directly from the European Standard.

RESULTS AND DISCUSSION

Preliminary experiments were conducted to establish which of the three methods of visualising voids for Image Analysis, outlined above, was best suited to the purpose.

The first method, low angle illumination, is most similar to the conventional method but it produces an image complex in both shape and grey level. (Fig. 2). Recognition and measurement of air-voids in this complex image would probably require software incorporating "artificial intelligence" and is beyond the scope of this work. This method was, therefore, discarded.

Pilot experiments with the other methods highlighted the key importance of surface preparation in producing an image for instrumental analysis. The data quality from the two methods was similar, and so the less laborious method of treating the surface with aluminium oxide powder was adopted as the method of choice (Figure 3).

Results obtained by Image Analysis are compared with those obtained by experienced operators using the linear traverse method. Data from both methods are recorded as a series of chord lengths from which computed void size distributions are calculated. The calculated distribution is based on a model which assumes only a nominal set of air void diameters are present. The nominal diameters are those corresponding to the maximum chord length in each of the stated size ranges. Of the other calculated parameters, in this paper we have placed most emphasis on those used for specifications, the spacing factor and the total air content.

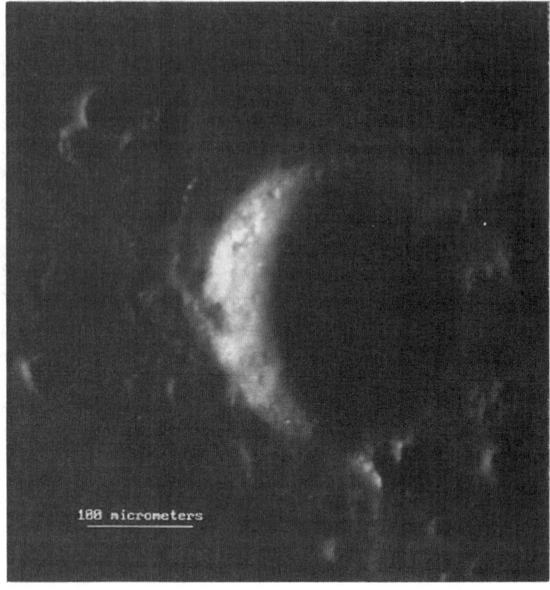

Fig. 2 Voids contrasted by low angle illumination

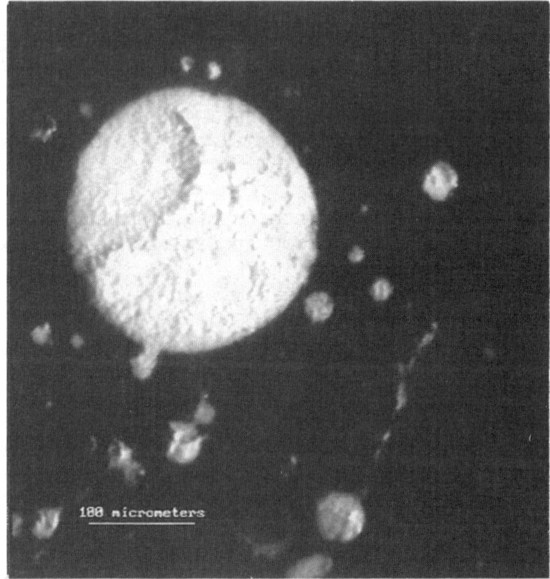

Fig. 3 Voids contrasted by filling with aluminium oxide powder

Plots of void size (diameter) distributions for the two methods are presented in figures 4 and 5. These data illustrate the similarities between the two methods at the upper end of the size range, whilst highlighting the significant differences below around 70 micrometers (μ), with the Image Analysis recording many more smaller voids.

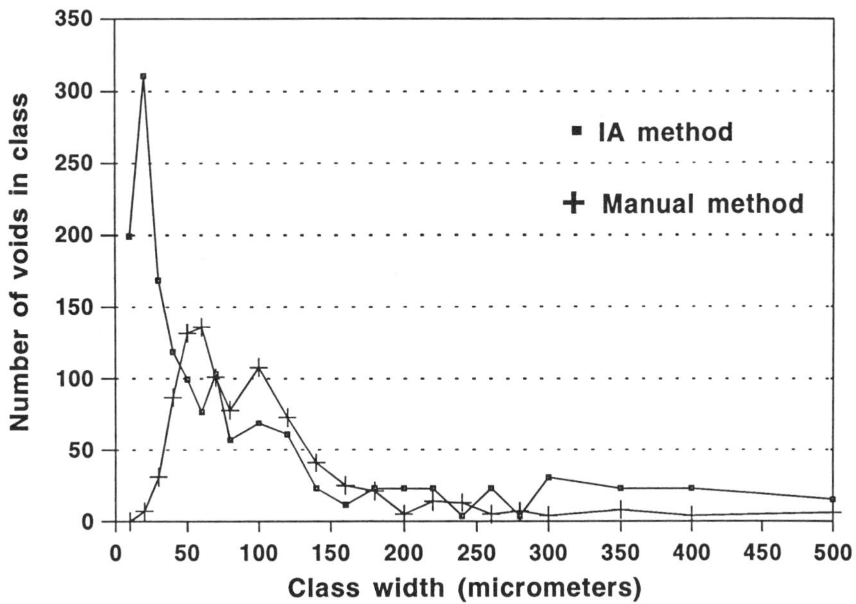

Figure 4 : Void Size Distribution from Stablair Air Entrainer

This major discrepancy between the two methods raises the question as to whether the smaller "voids" are artefacts due to surface roughness, or whether they are truly small air-voids which escape detection by the conventional procedure.

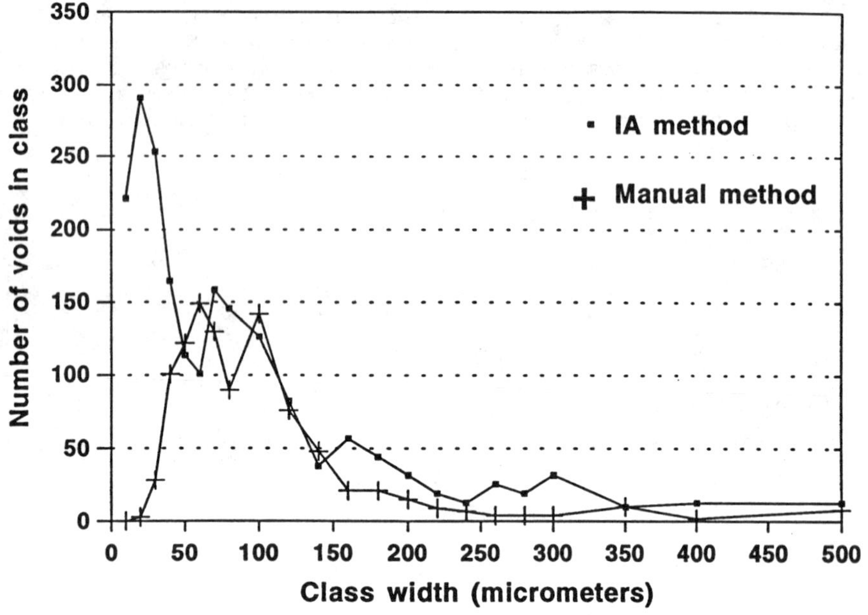

Figure 5 : Void Size Distribution from AE1 Air Entrainer

To investigate this further, a control sample containing no air entraining agent was prepared using aluminium oxide powder and measured using the Image Analysis technique. Again a large number of smaller "voids" were detected (Figure 6) indicating that "voids" detected in this region may largely be due to surface defects.

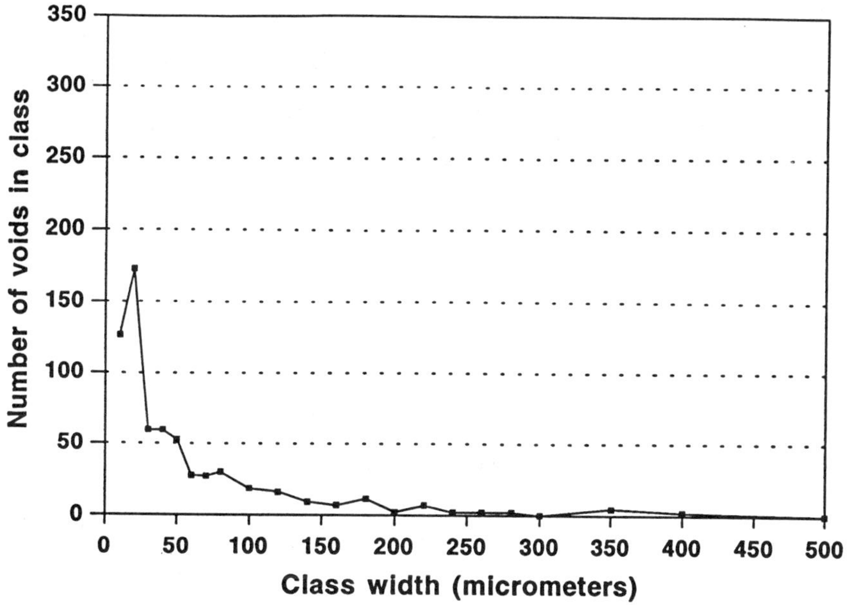

Figure 6 : Void Size Distribution of Control, by IA

The effect this discrepancy has on the parameters, % air and spacing factor are compared in Table 1.

Table 1 : Comparison of % Air and Spacing Factor for the manual and Image Analysis Techniques

TECHNIQUE	CORMIX AE1		CORMIX STABLAIR	
	% AIR	SPACING FACTOR µm	% AIR	SPACING FACTOR µm
Manual	4.2	130	4.0	140
Image Analysis	6.2	60	5.8	80
Image Analysis > 70 µ	4.9	110	4.8	120
Total Air Content Wet concrete	5.1		5.2	

The total air content as measured by Image Analysis is higher than that measured by the manual method. In addition, the very large number of small voids detected by Image Analysis greatly reduces the spacing factor. An arbitrary cut-off point selected to ignore the large number of small voids undetected by the manual method, results in a much improved approximation between the Image Analysis and manual techniques. For the current series of concretes, the optimum value was found to be around 70 micrometers.

Subsequent measurements for a range of air entraining agents in different concretes revealed the critical cut-off point varied significantly depending on the sample, leading the authors to believe that such a method was an oversimplification, and that a single cut-off point is unreliable.

A more rigorous technique using Electron Microscopy was selected to differentiate between surface defects and small air-voids.

A thin section of air entrained concrete was carbon coated in preparation for electron microscopy. An area of approximately 5 mm by 3 mm was imaged at a magnification with respect to the photo micrographs of 200x. In back scatter mode the electron microscope image revealed the subtleties of the three dimensional appearance of the air-voids. Identification by eye was easy, even if voids coalesce or were otherwise distorted to produce irregular shapes. However, these criteria for identification would be extremely complicated to programme into a computer whose capabilities are restricted to the recognition of simple geometrical shapes.

Since the magnification of the electron microscope is so great, the area which can be conveniently observed and hence the number of air-voids is relatively small. Every void in the area of the micrograph was therefore counted and measured, rather than attempting to reproduce the standard methods by following the traverse. In this way enough voids were observed to construct a size distribution (Figure 7).

Whilst this sample is not identical to those measured previously by the Image Analysis and the manual technique, two general features are noted.

1. The size distribution of air-voids as measured by electron microscopy is closer to that obtained by the manual technique than by Image Analysis.
2. Using a cut-off limit of say 70 micrometers is an oversimplification in that there are a significant number of true voids below this size.

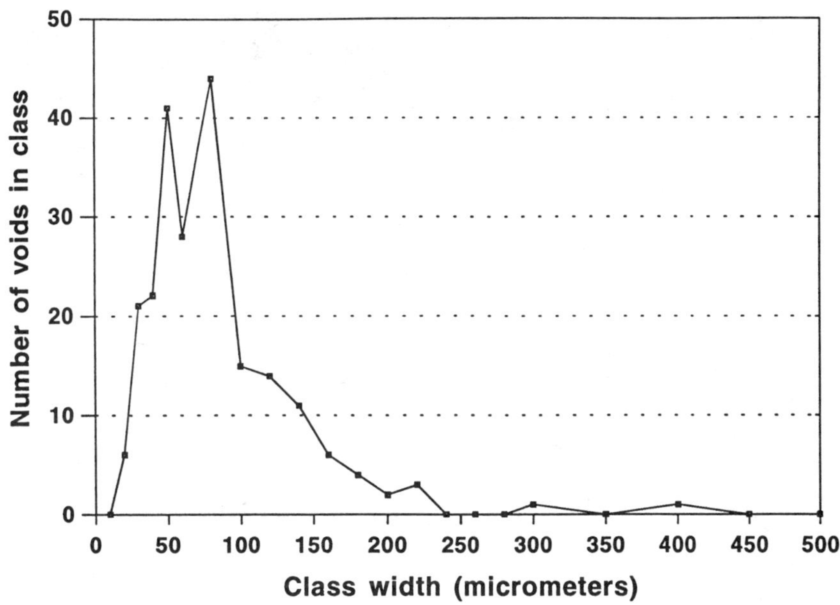

Figure 7 : Void Size Distribution of Air Entrained Concrete by Electron Microscopy.

The electron micrographs also highlight the importance of surface preparation for Image Analysis. (Figures 8 and 9). In Figure 8 the ease of identification of air-voids using the back scatter mode is illustrated. Figure 9 presents the same area of sample after treatment with zinc oxide powder. This figure clearly illustrates the problems associated with contrasting not only voids, but surface defects caused by porous aggregates and pull-out of sand and cement from both within the paste and around the void edges. The incorporation of a simple shape acceptance criteria was unable to eliminate this problem.

Although invaluable in resolving these questions, the use of the electron microscope is not suitable as a cost effective routine procedure.

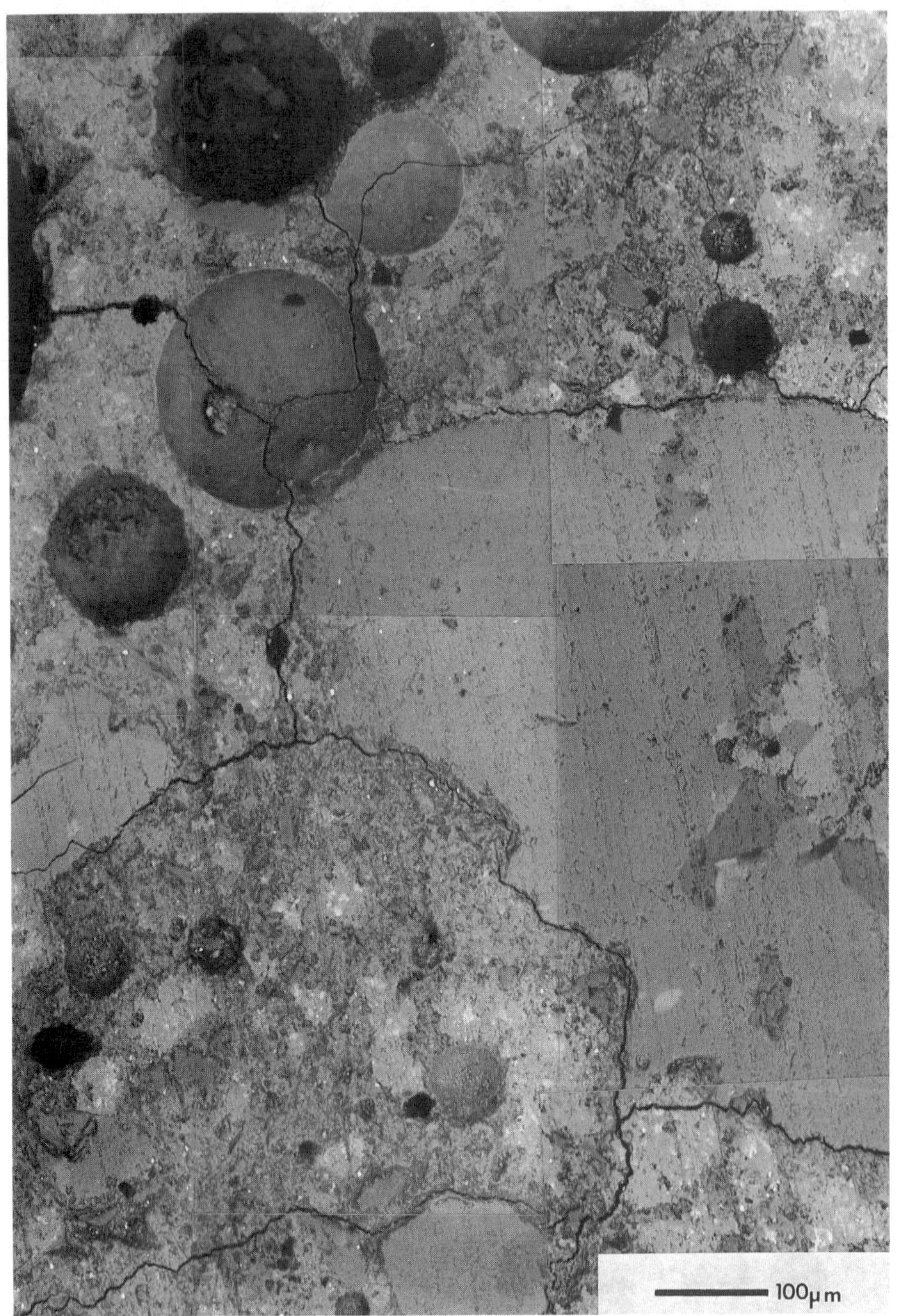

Figure 8 : Backscatter Electron Micrograph of Concrete.

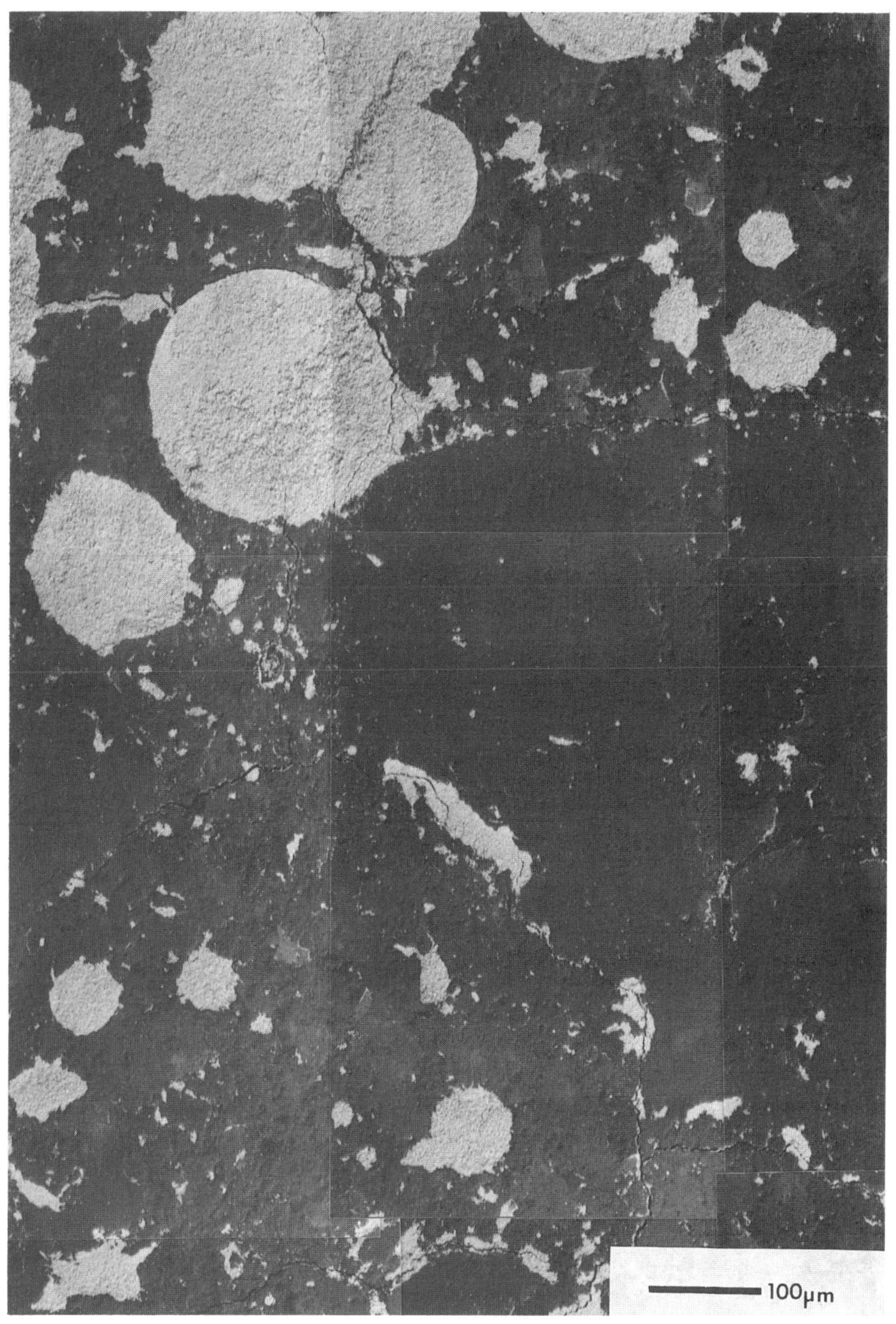

Figure 9 : Backscatter Electron Micrograph of Concrete Treated with Zinc Oxide Powder

CONCLUSIONS

The use of an Automated Image Analysis technique to measure air-void parameters of hardened concrete has been investigated.

The automatic technique developed can approximate the manual technique provided a suitable cut-off point is used to prevent surface defects being detected as small air voids.

The optimum cut-off point for different quality concretes has been found to vary, thus a single cut-off is considered an oversimplification.

Scanning electron microscopy has been used to differentiate between small air voids and surface defects. This technique highlights the problems associated with surface preparation for adequate visualisation of air voids by the automatic technique.

The void size distribution obtained from scanning electron microscopy has shown that the use of cut-off points to compensate for surface defects also eliminates some true air voids.

The automated technique developed overcomes the laborious, subjective measurements made by manual operators. It does not, however, improve the accuracy in its current form due to the approximation made when compensating for surface defects.

For the current automated Image Analysis system to be used as an accurate routine test procedure further developments are needed to improve the surface finish of samples minimising the errors attributed to surface defects. An alternative approach would be a more sophisticated "artificial intelligence" scanning the more complex images produced by low angle illuminations of the polished concrete surface.

ACKNOWLEDGEMENTS

The authors wish to thank Unilever Research, Port Sunlight Laboratories and Cormix Construction Chemicals for permission to publish this paper. Thanks are also extended to Mrs J Munro-Brown of Unilever Research for her contribution to the electron microscopy work.

References

[1] Recommended Practice for the Microscopial Determination of Air-Void Content and Other Parameters of the Air Void System in Hardened Concrete, ATTM C457-90. Annual Book of Standards, Vol 04.02, ASTM, Philadelphia, 1990.

[2] Efficancy Test for Air Entaining Admixtures. European Standard ENV 480 Part II for Determination of Air-void Characteristics in Hardened Concrete. 1991.

David R. Lankard,[1] Nick J. Scaglione,[1] and John E. Bennett[2]

PETROGRAPHIC EXAMINATION OF REINFORCED CONCRETE FROM CATHODICALLY PROTECTED STRUCTURES

REFERENCE: Lankard, D. R., Scaglione, N. J., and Bennett, J. E., **"Petrographic Examination of Reinforced Concrete from Cathodically Protected Structures,"** Petrography of Cementitious Materials, ASTM STP 1215, Sharon M. DeHayes and David Stark, Eds., American Society for Testing and Materials, Philadelphia, 1994.

ABSTRACT: Core samples were obtained from eleven 18-year to 32-year old reinforced concrete bridge and parking structures that have been under cathodic protection for 2 to 10 years. Characterization studies were conducted on the cores to assess the effects of the sustained cathodic protection current on the anode, the anode/concrete component, and the cathode/concrete component of the cathodic protection system. Characterization techniques included (1) conventional optical petrographic techniques, (2) SEM/EDS techniques, and (3) chloride ion content measurements. The 11 structures represent 8 different CP systems.

Data are also presented on cores taken from 3 ft. x 4 ft. x 6 in. (0.94 m x 1.22 m x 152 mm) reinforced concrete slabs powered for 30 months at 10 mA/ft^2, 20 mA/ft^2, and 40 mA/ft^2 (111 mA/m^2, 222 mA/m^2, and 444 mA/m^2) under laboratory conditions using four different CP anodes.

The investigation has provided new insights into the factors controlling the performance and longevity of the various cathodic protection systems.

KEYWORDS: Cathodic protection, corrosion, bridge decks, parking structures, chlorides, overlays

---

INTRODUCTION

With the advent of an increasing use of deicing salts on pavements and bridges in the United States in the early 1960's, a widespread corrosion-related problem began to occur at an increasing rate. In spite of the passivity provided to steel by portland cement concrete it has been shown that chloride ions contained in deicing salts, seawater, or admixtures can destroy the concrete's ability to

---

[1]President and Research Technician, respectively, Lankard Materials Laboratory, 400 Frank Road, Columbus, OH 43207

[2]Research Fellow, ELTECH Research Corporation, 625 East Street, Fairborn Harbor, OH 44077

keep the steel in a passive state. A chloride ion content in the range of 1.0 to 1.4 lb/yd^3 (0.59 kg to 1.95 kg/m^3) of concrete is the critical value above which steel corrosion in concrete can occur [1,2]. The resultant rust occupies more volume than the parent iron which exerts tensile stresses on the surrounding concrete. When the stresses exceed the tensile strength of the concrete, cracking develops. This cracking often interconnects between reinforcing bars and common undersurface fracture or delamination develops. As corrosion continues, the concrete cover breaks up and a pothole or spall is formed at a free surface, frequently accelerated by freeze/thaw cycling and traffic loadings.

A number of new materials and practices that have the potential for extending the service life of reinforced concrete structures subjected to outside chloride sources have been studied and implemented. However, many structures built prior to the mid-1980's remain salt-contaminated and continue to deteriorate.

Two possible solutions, both electrochemical, have been suggested for reinforced concrete structures that are already salt-contaminated and experiencing corrosion. These two solutions are (1) cathodic protection and (2) electrochemical chloride removal. Both of these techniques are possible since concrete is an ionic conductor and is capable of supporting a small flow of electric current.

Cathodic protection was first applied by R.F. Stratfull and co-workers at The California Department of Transportation on the Sly Park Road Bridge in June 1973 [3]. Many advances have subsequently been made in cathodic protection system components and installation procedures. Cathodic protection is now an accepted rehabilitation technique [4]. In a survey conducted in 1989, it was found that more than 275 bridge structures in the United States and Canada have been cathodically protected and that the total concrete surface under cathodic protection was about 9 million ft^2 (840,000 m^2) [5]. Additionally, a significant number of reinforced concrete parking structures have also been cathodically protected since the early 1980's.

The concept of cathodic protection as applied to reinforced concrete structures is relatively simple. It is recognized that the corrosion of steel in concrete is an anodic process. It follows that if the reinforcing steel can be induced to be more electronegative (cathodic), corrosion will be reduced. This is accomplished in practice by incorporating into the structure an auxiliary anode, where anodic reactions can take place without detriment. A relatively small flow of direct current between this auxiliary anode and the reinforcing steel is used to cause the steel to become more cathodic. The amount of current required is quite modest, about 1 mA/ft^2 (11 mA/m^2). Power consumption is on the order of 2-20 watts/1000 ft^2 (20-200 watts/1000 m^2). Cathodic protection, as applied to reinforced concrete, is a permanent installation and is intended to remain in place for the life of the structure.

In a large number of laboratory and field trials, cathodic protection of reinforced concrete members has provided dramatic reductions in the corrosion rate (and subsequent concrete deterioration). Despite this experience, however, the acceptance of cathodic protection for reinforced concrete structures has been relatively slow. Reasons for this have been the inherent difficulty of experimenting with steel in concrete and the complexity of concrete as an electrolyte. Chloride and moisture content can vary greatly throughout a structure. Such variations affect the electrical resistance and the distribution of current throughout a structure. Physical variations such as concrete cover and reinforcing bar spacing also affect the flow of electric current. Because of these complexities, the flow of current to the reinforcing steel is often unpredictable and may be locally inadequate or excessive. These factors contribute to the confusion and disagreement among workers regarding current control criteria. Also needed for improved user confidence in cathodic protection of reinforced concrete is proof that the treatment does not create significant adverse effects in the concrete, particularly in the vicinity of the anode and the cathode. In response to this latter concern, cores taken from cathodically protected reinforced concrete bridge and parking structures were examined petrographically.

CATHODIC PROTECTION SYSTEM OPERATION AND COMPONENTS

One of the principal differences in the various cathodic protection systems lies in the design and composition of the anode. Generically, anode materials fall into one of three categories including (1) coke asphalt, (2) conductive polymer, and (3) metallic.

Cathodic Protection Anodes

Coke asphalt--Coke asphalt is an asphaltic concrete that contains a high percentage of electronically conductive coke breeze. A typical coke asphalt mixture is,

Asphalt - 12%
Fine aggregate - 37%
Coarse aggregate - 10%
Coke breeze - 41%

In practice, the coke asphalt mixture is compacted onto the reinforced concrete wearing surface using the same equipment and procedures as used for conventional asphaltic concrete. Prior to placement of the coke asphalt material, primary anodes, in the form of cast iron discs, are placed on the reinforced concrete surface. After placement of the coke asphalt material at around a 2 in. (50.8 mm) thickness, it is overlaid with a 2 in. (50.8 mm) thickness of conventional asphalt concrete wearing surface.

Conductive polymers--Electrically conductive polymers are prepared by adding large quantities of carbon granules to an organic polymer matrix. These materials range from fine-grained coatings applied in thin [<1/32 in. - (0.8 mm)] layers aptly characterized as "conductive paints" to materials containing granules up to 1/8 in. (3.2 mm) in diameter, applied in thicknesses up to 1/2 in. (12.7 mm) or more which are more aptly characterized as "mortars".

Conductive polymers are used in a variety of compositions, consistencies, and geometric designs. In the cores examined here, conductive polymers had been used (1) as a conductive polymer overlay, (2) in a slot CP system, and (3) in a mounded CP system (see Fig. 1). In all of the CP systems involving conductive polymer anodes, the current is distributed by either metallic or graphite fiber strands while the conductive polymer serves as the surface where anode reactions occur.

A commercially manufactured anode is a copper wire encapsulated with a relatively thick layer of conductive polymer.

Metallic anodes--Metallic anodes used in cathodic protection systems include catalyzed titanium and zinc. Titanium, containing a catalytic coating, is used in both ribbon and mesh form (both commercially manufactured).

Zinc is applied to concrete surfaces as a flame-sprayed or arc-sprayed coating or layer. Of all of the anodes used, zinc is the only material that is comsumed and may function to a degree without the need for impressed current.

Cathodic Protection Events

Figure 1 is a schematic diagram [6] showing the components of an impressed current cathodic protection system for reinforced concrete. Also shown is the direction of current flow.

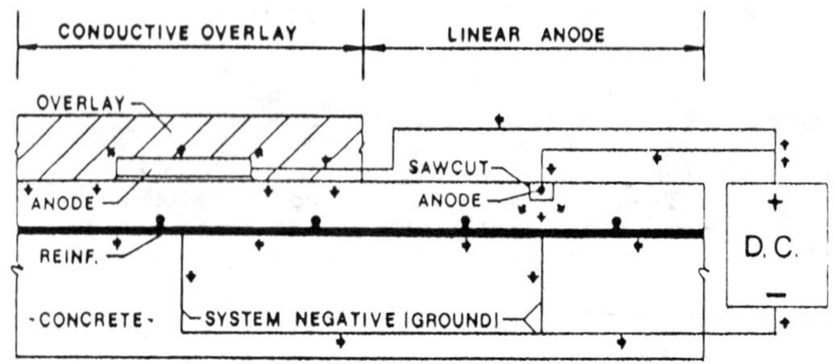

FIG. 1--Schematic diagram showing the various components of a cathodic protection system on a reinforced concrete structure and showing the direction of positive current flow in the systems (Reference 6).

During the course of an impressed current cathodic protection treatment, a number of events occur that can have consequences regarding the physical and chemical integrity of the various components of the CP system. These events include:

1. The migration of anions and cations in the concrete pore water toward the anode and cathode, respectively.
2. Electrochemical reactions at the anode/concrete interface.
3. Electrochemical reactions at the cathode/concrete interface.

The movement of ions through concrete under the influence of an electrical field is well established [7,8]. It is obvious that ionic movement can occur in concrete even under the relatively low potential differences characteristic of cathodic protection. Specifically, in a cathodic protection environment, Ca^{++}, Na^+, and K^+ can migrate toward the cathode (reinforcing steel) while the OH^- and Cl^- can migrate toward the anode. In practice, it appears that K^+ has the greatest mobility among the available cation species, while Cl^- is the most mobile anion. Associated with these events is the possibility that an increased concentration of alkali cations near the cathode may initiate alkali-aggregate reactions under conditions which prior to the CP treatment were innocuous.

The potentially deleterious effects of anodic reactions on concrete has long been recognized within the cathodic protection industry. When current passes from an anode into an electrolyte (i.e., concrete), electrochemical reactions occur at the surface of that anode. In the case of chloride contaminated concrete, two reactions are possible when using inert anodes:

(1) $2H_2O \rightarrow O_2 + 4H^+ + 4e^-$ or
(2) $2Cl^- \rightarrow Cl_2 + 2e^-$

In the presence of water and a pH greater than 4, the latter reaction will be followed by rapid hydrolysis of chloride as follows:

(3) $Cl_2 + H_2O \rightarrow ClO^- + Cl^- + 2H^+$

In either case, one mole of acid (H^+) will be generated for each Faraday of current passed (26.8 amp-hr).

When carbon-based anodes are used, a small amount of current will be expended for the oxidation of carbon as follow:

$$(4) \quad C + H_2O \rightarrow CO + 2H^+ + 2e^- \quad \text{or}$$
$$(5) \quad C + 2H_2O \rightarrow CO_2 + 4H^+ + 4e^-$$

In the presence of water and a pH greater than 10, carbon dioxide will be converted to carbonate, as follows:

$$(6) \quad CO_2 + H_2O \rightarrow CO_3^{-2} + 2H^+$$

In each case, the passage of current from the non-sacrificial anode to the cathode generates acid. The cement hydrate phases in concrete are vulnerable to acid attack. Carbonate aggregates (limestones/dolomites) are also vulnerable to acid attack.

Under normal conditions, the electrochemical reaction occurring at the cathode is:

$$(7) \quad 2H_2O + O_2 + 4e^- \rightarrow 4OH^-$$

At high electrical potentials, molecular hydrogen can be generated at the cathode via the following reaction:

$$(8) \quad 2H^+ + 2e^- \rightarrow H_2$$

This latter reaction is of concern regarding the phenomenon of hydrogen embrittlement of the reinforcing steel.

Concern has also been expressed regarding the effect of the ionic migration on the bond strength of the reinforcing steel [6].

In summary, the potential adverse effects of cathodic protection treatments on the various components of the cathodic protection system include:

1. Acid attack of concrete surrounding and adjacent to the anode (principally manifested as a dissolution/softening of the hydrated cement phases and carbonate aggregate phases). Degradation of the anode itself is also a possibility.
2. Initiation of alkali-silica reactions in concrete surrounding and adjacent to the cathode (manifested as both a softening and cracking adjacent to the reacting particles).
3. Chemical changes in the concrete at the cathode/concrete interface resulting in a reduction in bond strength.

PETROGRAPHIC EXAMINATIONS

Forty-five concrete cores from reinforced concrete members currently under cathodic protection were examined. The cores represent nine different bridge structures, two parking structures, and large concrete slabs used in a controlled test plot experiment. The CP systems have been operating for various periods of time ranging from 2 years to 10 years.

The examination of the concrete cores was intended to determine the effect of the cathodic protection treatment on (1) the anode, (2) the concrete surrounding the anode, (3) the bond between the overlay concrete (where applicable) and the base slab concrete, (4) the cathode (reinforcing steel), and (5) concrete surrounding the cathode.

All of the cores were initially subjected to a preliminary characterization which included:

1. Dimensional measurements of core diameter and length and dimensional measurements identifying locations of anode and cathode components.
2. Visual examination (including examination under the stereomicroscope) to identify obvious distress features such as cracking, delaminations, discoloration, and softening on cored surfaces and end surfaces of the cores.
3. Photographic documentation of the features identified in (1) and (2).

The cores were subjected to more detailed analyses which included:

1. Destructive examination of the anode/concrete component and the cathode/concrete component of the system. This work was done on existing fracture surfaces, fresh fracture surfaces, and on lapped surfaces. Characterization procedures included reflected light microscopy and scanning electron microscopy. Measurements of pH were made on fresh fracture surfaces and fresh contact surfaces using color-sensitive indicator solutions*.
2. Destructive examination of the concrete located between the cathode and anode (on fresh fracture surfaces and on lapped surfaces). Characterization procedures used here included reflected light microscopy, scanning electron microscopy, energy dispersive x-ray spectroscopy, and ultraviolet fluorescence [9].
3. Characterization of the concrete representing the original structure including assessment of overall quality, identification of aggregate types, estimates of cement content and water-cement ratio, assessment of the quality of the entrained air system, and degree of consolidation.
4. In some core samples, chloride ion content measurements were made on powdered samples obtained at selected elevations as drill debris in accordance with AASHTO Designation T260-82, The Standard Method of Test For Sampling and Testing For Total Chloride Ion in Concrete.

RESULTS

Anode Performance

The performance of the anode materials was assessed on the basis of their ability to distribute electric current without undergoing deleterious breakdown such as cracking, softening, or dissolution.

Coke asphalt--Three of the bridge structure concretes examined have coke asphalt overlay anodes. In the cores examined, only the coke asphalt layer itself was available for examination. For this system, cores were not taken through the primary current distributor (sometimes called the primary anode). The anodes had been in operation for 4 to 5 years. Performance assessment of the coke asphalt anodes was made on the basis of (1) the condition of the coke asphalt layer bond to the substrate concrete, (2) adequacy of compaction of the coke asphalt layer, (3) observations of cracking in the coke asphalt layer, and (4) evidence of softening, dissolution, or discoloration.

Based on these performance factors, it is judged that the coke asphalt anode layers on all three structures was in good condition following the 4 to 5 year service period. At the time the coring was done, the coke asphalt layer was bonded to both the asphalt overlay and the original portland cement concrete bridge deck wearing surface.

---

*Rainbow Indicator, Germann Instruments, Inc., Chicago, Illinois

One deficiency in the coke asphalt layer at all three installations is substandard compaction which has resulted in an abnormally high amount of interconnected voids. While this has not yet adversely affected the durability of the anode, it would be of interest to learn the effect of this variable on the efficiency of distribution of the cathodic protection current.

Conductive polymer in slots--The slot anode system was used on two of the structures (bridge decks) evaluated here. These CP systems had operated for 6 and 9 years. Assessment of the performance of these anodes was made on the basis of observations of (1) the condition of the bond of the conductive polymer to concrete in the slot, (2) condition of bond of the conductive polymer to the primary current carrier (graphite fiber strand), (3) observations of cracking in the conductive polymer, and (4) observations of softening, dissolution, or discoloration of the polymer.

At the site where the CP system has been operating for 9 years, the conductive polymer shows virtually no chemical alteration and only minor cracking. Cracking within the conductive polymer, where it does occur, is principally through the matrix phase passing around the carbon granules. Cracking occurs within those regions that are somewhat deficient in carbon granules (rich in polymer matrix phase). There is minor disbonding of the conductive polymer from the concrete along portions of the anode slot.

At three of the four coring sites on the structure where the CP system has been operative for 6 years, the conductive polymer is also in very good condition. At one of these sites, however, there has been a major breakdown in the conductive polymer anode in the form of cracking, spalling, and dissolution. The dissolution in this case primarily involves consumption of all or portions of carbon granules within the polymer matrix.

Conductive polymer mounds--Cores were taken from three different CP systems in which conductive polymer mounds served as a distribution anode. These included a bridge deck (9 years of operation), a parking structure (8 years of operation) and a large reinforced concrete slab (operated for 30 months under controlled laboratory conditions).

Assessment of the performance of the mounded conductive polymer anodes was done on the same basis as used for the conductive polymer anodes in slots.

In the bridge deck system, the conductive polymer mounds are in excellent condition following the 9 years of service. The polymer is well compacted and shows no significant chemical alteration. A tight bond persists between the anode mound and the bridge deck wearing surface. There is minor disruption of the bond between the anode and the overlay concrete. This disbonding (hairline cracking) typically occurs along a free surface of the conductive polymer material at locations where the concentration of air voids is higher than normal (i.e., along the top free surface of the polymer mound).

Some degradation of the conductive polymer anode mound has occurred in all six cores taken from the parking structure deck (8 years of operation). The degradation is characterized as a dissolution/discoloration of the polymer matrix phase and a dissolution (loss of mass) of carbon granules. In general, the amount of anode that has degraded is less in the cores taken from zones of more moderate circuit voltage. Where degradation of the anode has occurred in all of the cores, it is characterized as relatively mild. The degraded regions of the anode are still quite competent, the polymer still forms a continuous phase, and carbon granules (while somewhat reduced in size) are still intact.

The laboratory slab was operated at a current density of 20 mA/ft^2 (222 mA/m^2) for a 30 month period. This is four times the maximum recommended current density for this anode material. Following this treatment, the conductive polymer anode was generally in good condition, showing no cracking or significant chemical alteration. In the cores examined from this site, the polymer was not fully consolidated, showing a significant amount of fine porosity and a slight amount of honeycomb. This porosity in the anode material had been intruded by cement paste material which contained high levels of both chlorine and sulfur.

Conductive polymer overlay--The conductive polymer overlay CP system was used on one of the structures (bridge deck) evaluated here. This CP system was installed in 1988. The current distributor to the anode is a welded steel wire fabric and the active anode surface is the conductive polymer overlay. The conductive polymer is a mixture of pea gravel, coke breeze, and polyester resin. It is 1 in. (25.4 mm) to 1-1/2 in. (38.1 mm) thick. At the time of coring, this CP system had been in operation for 4 years.

The conductive polymer overlay exhibited the most pronounced deterioration. Distress in the overlay is associated with corrosion (oxidation) of the steel welded wire fabric. This phenomenon is illustrated in Figure 2 which shows section views (perpendicular to the wearing surface) of cores taken from this bridge deck. Cracking in the polymer concrete overlay emanates from the corroding steel. Cracking distress in the polymer overlay was observed at two of the four coring sites examined here.

Arc-sprayed zinc--Arc-sprayed zinc coatings were used as the anode layer in two of the structures evaluated here including piers on a bridge in contact with seawater and a bridge deck. The former installation has been in operation for about 1-1/2 years and the latter about 4 years. Factors used in the performance assessment of the zinc anodes included observations of (1) bond of zinc to the concrete and (2) observations of cracking, softening, discoloration, or dissolution of the anode.

Overall, the zinc anodes are in good condition following 1-1/2 to 4 years of service at these installations. The only alteration of the zinc is the formation of relatively small amounts of zinc oxide on free surfaces. On the bridge piers, the effect is primarily cosmetic. On the bridge deck, the oxidation of the zinc anode may contribute to a reduction in the quality of the bond between the zinc anode and the bridge deck concrete.

At both installations, the zinc layer shows a considerable amount of internal porosity. This feature may be characteristic of arc-sprayed metallic coatings. However, the effect of this variable on anode efficiency should be considered.

Titanium--Titanium mesh or ribbon anodes were used on four of the structures evaluated here. They included (1) titanium ribbon on a bridge deck installed in 1990, (2) titanium mesh on a bridge deck installed 1986, (3) titanium ribbon on a parking structure deck installed in 1985, and (4) ribbon and mesh in test slabs operated for 30 months under laboratory conditions. In the latter installation, anode current density was 10, 20, and 40 mA/ft^2 (111, 222, and 444 mA/m^2). The current densities of 20 mA/ft^2 and 40 mA/ft^2 (222 mA/m^2 and 444 mA/m^2) represent two and four times the maximum recommended current density for this material.

At all of these sites, the titanium anodes (mesh and ribbon) are in excellent condition and show no alteration or distress.

Copper wire/conductive polymer--This anode is a 0.06 in. (1.52 mm) diameter copper wire surrounded by a 0.33 in. (8.38 mm) diameter conductive polymer encapsulant. This anode was used in the CP system on one of the laboratory test slabs operated for 30 months at 20 mA/ft^2 (222 mA/m^2). This current density is about twice the maximum recommended for this anode material.

Overlay Wearing Surface

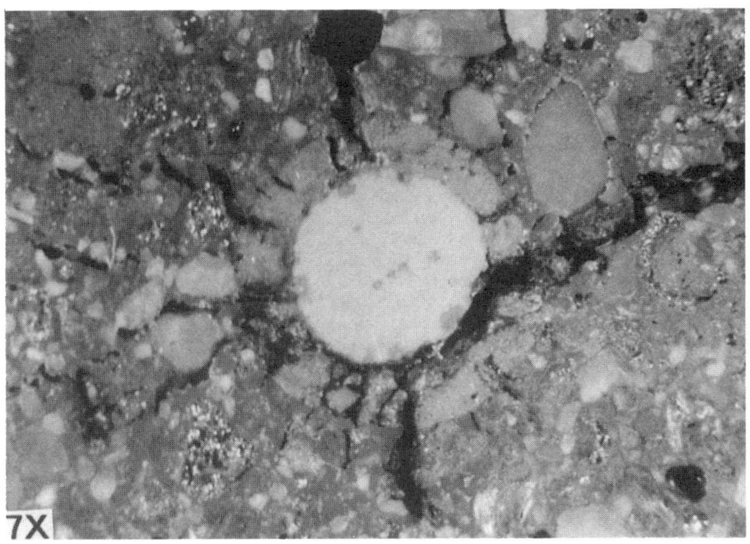

FIG. 2--Cracking in a conductive polymer overlay associated with corrosion of the steel mesh current distributor.

On a gross scale, this anode appears to have been relatively unaffected by the 30 month CP treatment. However, it is evident that there has been some reaction between the anode surface and the surrounding concrete. A black skin formed on the concrete at the interface was derived from the anode. Some copper has migrated through the polymer to the adjacent concrete.

Concrete Performance Assessment

For all of the cathodic protection systems evaluated here, portland cement concrete is the electrolyte (ionic conductor) that electrically ties together the anode and the cathode (reinforcing steel) in the system.

The three elements to consider regarding the performance and role of the concrete in a cathodic protection system include:

1. The concrete/anode component (i.e., concrete in direct contact with the anode).
2. The concrete/cathode component [i.e., concrete in direct contact with the cathode (rebar)].
3. The concrete lying between the cathode and the anode.

All three of these components were carefully examined in the present investigation to learn the effect of the CP system design and the concrete composition on their stability.

Concrete type and composition--All portland cement concretes examined in the present investigation include the original concrete in the structure and, in some cases, concretes used for repair and for overlays. Twenty different portland cement concrete compositions were examined. Widely different aggregate types and cement contents ranging from 540 lb/yd^3 (322 kg/m^3) to over 800 lb/yd^3 (478 kg/m^3) were represented. In addition to conventional air-entrained and non-air-entrained portland cement concretes, superplasticized dense concretes, and latex-modified concretes in contact with the anodes were also represented.

Despite the fact that the concretes came from many different bridge and parking structures in 8 U.S. states and one Canadian province, they all had one feature in common (e.g., they are all good quality concretes having low to moderately low water-cement ratios. Samples of concrete that were air-entrained had good entrained air void systems. Aggregates ran the gamut of lithologies and, in a number of cases, included rock types that are potentially alkali-silica reactive (including rhyolite, argillite, chert, gneiss, greywacke, basalt, and quartzite). In almost all cases, the aggregates used in the portland cement concretes in these structures have, over the 20 to 30 year service life of the structures, shown good durability.

Concrete/cathode component--For all of the cathodic protection systems considered here, the concrete/cathode component of the system does not represent a problem area. In no case has concrete in direct contact with the reinforcing steel experienced any significant softening, dissolution, discoloration, or other forms of chemical alteration. Typically, hydrated cement paste in direct contact with the rebars has the same hardness, color, and pH level as cement paste at sites well away from the reinforcing steel.

In a few instances, some light to moderate cracking in the concrete surrounding the rebar was observed. In these cases, the evidence suggests that the origin of the cracking is either structural or corrosion-related and is not a consequence of the effects of the sustained cathodic protection current.

Of the 45 cores examined here, evidence of alkali-silica reactions was found in only 5. All 5 cores came from the same bridge in Minnesota (3 from the deck and 2 from the sidewalk). In these cores, a relatively few chert aggregate particles (from the fine aggregate fraction) showed evidence of alkali-silica reactions. These reacted particles occurred both above and below the top rebar (cathode) and were not concentrated in the concrete lying directly over the top rebar. From this evidence, it was concluded that the alkali-silica reaction was previously present in this concrete and was not initiated by the CP treatment. In any event, the distress is confined to the reacted particles (reaction rims and internal particle cracking); concrete adjacent to reacted particles shows no cracking or distress.

Concrete/anode component--In virtually all of the cores examined here (45) there was some alteration and/or cracking of concrete in direct contact with the anode. This degradation ranged from an insignificant discoloration on the one hand to total destruction of the concrete on the other.

The alteration can be characterized principally as a chemical attack of the hydrated portland cement phases by an aqueous acid solution. In several instances, limestone aggregate particles adjacent to the anode were also affected. The consequence of the acid attack is a softening of the concrete, and in the case of the hydrated cement phases, a significant reduction in pH levels (from the normal 12.5 to as low as 5 to 7).

Somewhat less of a problem is minor cracking in concrete adjacent to the anode that can be directly attributed to the effect of the sustained cathodic protection current.

In Table 1, an effort is made to quantify the extent of chemical alteration and cracking of concrete in direct contact with the anode for the various cathodic protection systems evaluated here.

The data presented in Table 1 should be viewed with the knowledge that each anode system evaluated is represented by a different number of cores and CP installation sites. It is not known with certainty how representative the cores are of the entire structure. Finally, only in the case of the laboratory conducted test plot is the current density and total charge passed known.

Despite these limitations, a significant observation is that for anodes having the potential of more uniformly distributing the cathodic protection current, the deterioration of concrete in contact with the anode is typically minimal. This includes (1) the coke asphalt overlays, (2) the conductive polymer overlay, and (3) the titanium mesh. Additionally, concrete in direct contact with the zinc coating (not an impressed current system) also showed virtually no distress.

Examples of alteration of concrete in direct contact with the various CP anodes are shown in Figures 3 through 6.

Figure 3 shows sections views (perpendicular to the wearing surface) of as-cored surfaces of a core taken from a parking structure deck currently under cathodic protection. The parking structure was constructed in the early 1970's and a mounded conductive polymer CP system was installed in 1984. The top photograph shows a conductive polymer anode mound between the latex-modified concrete overlay (top) and the original deck concrete (bottom). Hydrated cement paste phases in both concretes in contact with the anode have been degraded by acid attack with the latex-modified concrete showing the greatest extent of deterioration. A vertically-oriented crack passes through both concretes and the anode. The bottom photograph in Figure 3 shows acid attack of a limestone aggregate particle in the overlay concrete adjacent to the anode.

Figure 4 is a core taken from a cathodically protected parking structure (deck) built in the early 1970's. A titanium ribbon anode CP system was installed in the mid-1980's. The ribbon anode is completely encapsulated by latex-modified concrete containing a trap rock coarse aggregate and a natural quartz sand. The hydrated cement paste phase in the latex-modified concrete in direct contact with the

TABLE 1--Assessment of extent of chemical alteration and cracking in concrete in direct contact with various cathodic protection anodes.

Anode System	Chemical Alteration of Concrete in Contact With the Anode				Cracking Attributed to the Cathodic Protection Treatment in Concrete in Contact With the Anode			
	None	Light	Moderate	Heavy	None	Light	Moderate	Heavy
Coke Asphalt	L	T	...	...	T	...	...	...
Conductive Polymer in Slots	...	T	L	...	...	T	...	L
Conductive Polymer Mounds	...	T	...	L	...	T	...	...
Conductive Polymer Overlay	T	...	...	...	T	...	...	...
Zinc	T	...	...	...	T	...	...	...
Titanium Mesh	T	L	...	...	T	...	...	...
Titanium Ribbon	...	...	T	L	...	T	...	...
Copper/Polymer	...	...	T	...	...	...	T	...

T = Typical
L = Less Common

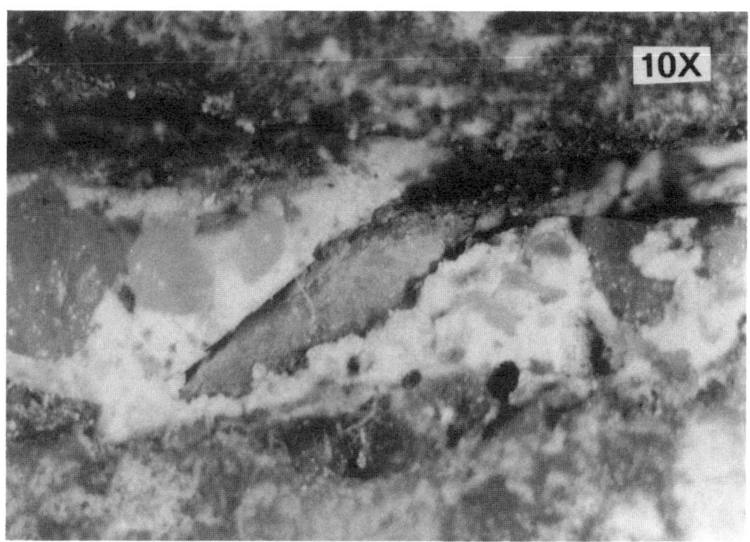

FIG. 3--Acid attack and cracking in concrete adjacent to a conductive polymer mound anode.

FIG. 4--Chemical attack and cracking in latex-modified concrete from a titanium ribbon anode CP system.

anode has undergone significant acid attack. At this coring site, there is only 3/4 in. (19 mm) of concrete between the anode and the cathode. Cracks oriented perpendicular to the wearing surface pass through the level of both the anode and the cathode.

Figure 5 shows section views (perpendicular to the wearing surface) of a core taken from a cathodically protected bridge deck (and sidewalk) constructed in 1964. A slotted conductive polymer CP system was installed on the sidewalk in 1983. The conductive polymer is in good condition showing only minor cracking near the interface with the concrete and partial separation at the concrete/anode interface. Degradation of concrete in contact with the anode is limited to very minor cracking and insignificant softening.

There was virtually no degradation of concrete in contact with coke asphalt current distribution materials. Figure 6 is a section view (perpendicular to the wearing surface) of a concrete core taken from a bridge deck currently under cathodic protection with a coke asphalt CP system. The CP system was installed on the 26-year old structure in 1987. The coke asphalt layer in contact with the portland cement concrete serves as the anode and is in excellent condition after 5 years of service. Concrete in contact with the distribution anode shows no evidence of acid attack.

Figure 7 shows a core taken from one of the test slabs following a 30 month CP treatment using a conductive polymer encapsulated copper wire as the CP anode. This test slab was operated at 20 mA/ft^2 (222 mA/m^2) which is twice the current density normally used with this anode. At this coring site, the superplasticized dense concrete overlay containing the anode completely delaminated from the base slab concrete. Cracking in the overlay concrete, emanating from the anode, strongly implicates the anode as playing a major role in the delamination of the overlay. Overlay concrete in contact with the anode exhibits discoloration, softening, and cracking. Both hydrated cement paste phases and limestone aggregate particles in the concrete show evidence of alteration.

In the laboratory-controlled test plot experiment, current density was studied as a variable in three slabs containing the titanium mesh anode. Three levels of current density were used including 10 mA/ft^2, 20 mA/ft^2, and 40 mA/ft^2 (111 mA/m^2, 222 mA/m^2, and 444 mA/m^2). The maximum recommended current density at the surface of this anode is 10 mA/ft^2 (111 mA/m^2). The most significant effect of the variation in current density was an increase in the extent of alteration of the concrete in contact with the anode (superplasticized dense concrete). The greatest amount of alteration of concrete occurred at 40 mA/ft^2 (444 mA/m^2). Figure 8 shows a fracture plane through the titanium mesh anode from the slab operated at 40 mA/ft^2 (444 mA/m^2). Cement paste in direct contact with the anode shows discoloration and softening as a result of acid attack. The color change in the hydrated portland cement phases is from the normal grey to a lighter grey, and in cases of severe alteration, to orange and white. The alteration is limited to the 1 mm or so thickness of concrete in immediate contact with the anode. Acid attack is greatest near the junction of two anode strands as shown in Figure 8. At the lower current densities, alteration of concrete in contact with the mesh anode was significantly less and there was very little difference in the extent of alteration at 10 mA/ft^2 and 20 mA/ft^2 (111 mA/m^2 and 222 mA/m^2).

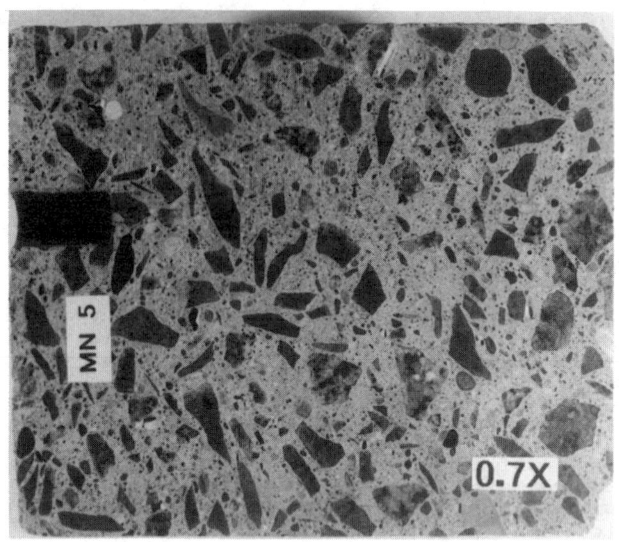

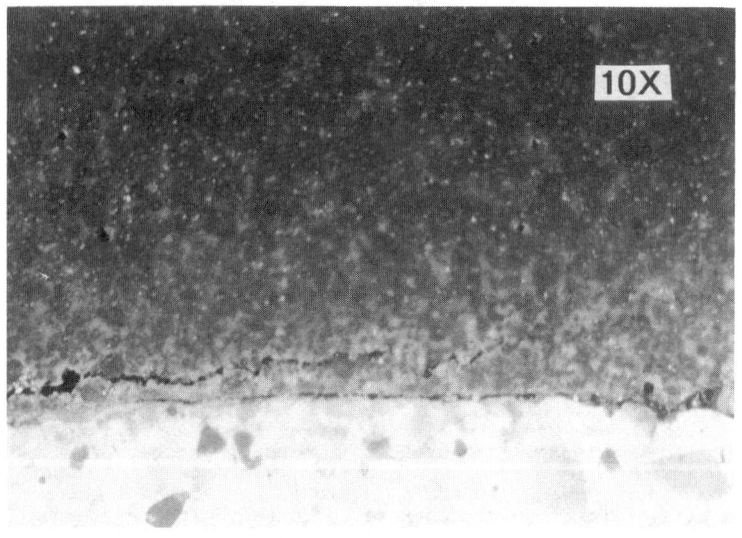

FIG. 5--Minor cracking in the anode and concrete from a slotted conductive polymer CP system.

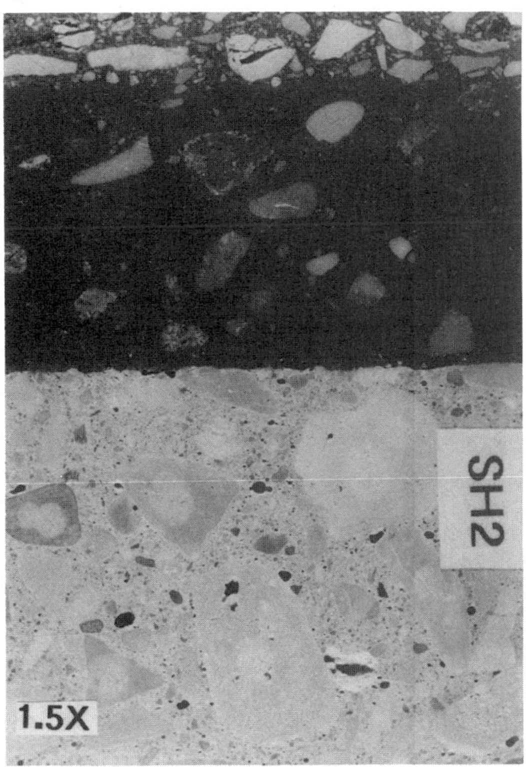

FIG. 6--Coke asphalt layer in contact with portland cement concrete from a bridge deck under cathodic protection for 5 years.

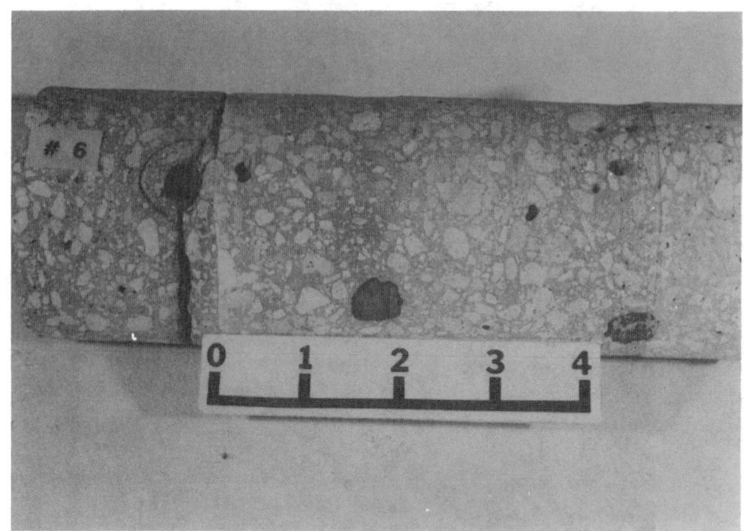

FIG. 7--Cracking and alteration of concrete encapsulating a conductive polymer/copper wire CP anode.

FIG. 8--Acid attack of superplasticized dense concrete in contact with a titanium mesh anode.

Concrete lying between the cathode and anode--For the 45 cores examined here, the thickness of concrete lying between the cathode and anode varied from a low of 1/4 in. (6.35 mm) to a high of 4-3/4 in. (12.1 cm).

Typically, concrete lying between the cathode and anode showed no significant cracking. In these cases, the concrete layer exhibited no significant chemical alteration attributed to the effects of the cathodic protection treatment (i.e., softening, discoloration, dissolution).

In those instances where significant levels of cracking occurred, it was difficult to judge whether or not the cracking was a consequence of the effects of the cathodic protection treatment. Where vertically oriented cracks provided direct communication between the anode and cathode, significant deterioration of the concrete occurred in concrete adjacent to the fracture surfaces.

Overlay Performance Assessment

In a majority of the cathodic protection systems evaluated here, the reinforced concrete member under protection is a bridge deck or parking structure deck. In the majority of these examples, an overlay was used to protect the cathodic protection anode and to provide a suitable wearing surface for the decks.

For coke asphalt and zinc strip anodes, an asphalt overlay is used. For mounded conductive polymer anodes and titanium anodes, portland cement concrete (including latex-modified concrete and superplasticized dense concrete) overlays are used. In one of the systems, a conductive polymer anode also functions as the wearing surface overlay.

In those instances where the anode itself is the overlay and where the anode is partially or completely encapsulated by the overlay material, the consequences of debonding of the overlay must be considered as it affects the efficiency of current distribution.

Many of the cores examined here which contained an overlay showed full or partial debonding of the overlay from the substrate concrete. Because of the nature of this investigation, it is not possible to say with certainty what role the cathodic protection treatment played in the debonding phenomenon. It is well known that debonding of overlays on concrete decks is a very complex issue.

Again, despite this limitation, a significant observation in the present investigation is the fact that CP systems incorporating anodes with a potential of uniformly spreading the cathodic protection current show less proneness to debonding of the overlay component. These include the coke asphalt overlays, the polymer concrete overlay, and the titanium mesh (the latter contained in a superplasticized dense concrete in two instances).

CONCLUSIONS

Forty-five concrete cores, taken from cathodically protected reinforced concrete structures were examined petrographically to learn the effects of the sustained cathodic protection currents on the durability of the various cathodic protection system components. The majority of cores examined (31) were taken from reinforced concrete bridges (typically bridge decks) that have been under CP protection for 2 to 10 years. Cores were also obtained from two parking structures under CP protection for 4 to 5 years and from a laboratory controlled field plot where cathodic protection current densities of 10, 20, and 40 mA/ft^2 (111, 222, and 444 mA/m^2) were maintained for 30 months. The examination provided some insights into the relative behavior of the various CP system components and the factors affecting this behavior.

Within the limitations imposed by the conditions of this investigation (sampling biases, limited and variable system operation period, lack of data on current density at the specific sampling sites), the following conclusions and salient observations are offered.

Anode Performance

1. Titanium anodes (mesh and ribbon) which contain a catalytic oxide coating have shown virtually no alteration or degradation.
2. Arc-sprayed zinc coatings and strips have shown only minor alteration in the form of oxidation of the zinc on free surfaces. In one instance, the formation of this oxide layer may have adversely affected the bond of the zinc to the concrete.
3. Coke asphalt overlays which function as secondary distribution anodes have generally performed well showing no significant alteration or degradation and maintaining a bond to the portland cement concrete under cathodic protection. Cores available for examination did not contain the anode current distributor (typically cast iron).
4. Conductive polymer anodes have generally shown good performance but, in some instances, can experience degradation in the form of cracking, dissolution, and softening.

5. Copper wire anodes encapsulated by a conductive polymer showed light degradation of the conductive polymer and migration of copper through the polymer to the concrete.

Concrete/Anode Component

1. It is a principal finding of the present investigation that, by far, the greatest potential "problem" area is the concrete/anode component of cathodic protection systems.
2. The problem is associated with the formation of acid by anodic reactions. The acid attacks the hydrated portland cement phases and adjacent carbonate aggregate particles (limestones/dolomites).
3. Some acid attack of the concrete in direct contact with the anode was observed in virtually all of the cores examined here (45). The severity of the attack ranged from insignificant (very minor softening of a thin (<0.1 mm) layer of paste to virtually complete degradation of a 1/8 in. to 1/4 in. (3.2 mm to 6.4 mm) layer of concrete in direct contact with the anode (a total loss of binding quality in the cement paste phase).
4. Acid attack of concrete in direct contact with the anode was least severe with those anodes most capable of uniformly distributing the cathodic protection current over the substrate concrete surface (including coke asphalt overlays, a conductive polymer overlay, and titanium mesh).
5. The consequences of the chemical degradation of concrete in contact with the anode need to be assessed. This event has the potential for disrupting the continuity of the electrolyte (ionically conducting concrete) between the cathode and anode. However, for the systems evaluated in this study, no system failure has yet occurred for this reason.
6. Future consideration should be given to the use of more acid-resistant concretes to encapsulate metallic and conductive polymer anodes. The use of high quality concretes with low water-cement ratios (e.g., superplasticized dense concrete/latex-modified concrete) does not guarantee that concrete in contact with the anode will not be attacked. Researchers may want to consider calcium aluminate cements, alkali-silicate cements, or magnesium phosphate cements for this application.

Concrete/Cathode Component

1. For the cathodic protection systems evaluated, the concrete/cathode component of the system is not a problem area. In all of the cores examined here (45), concrete in direct contact with the cathode (rebar) showed the same physical and chemical characteristics as concrete at sites well removed from the rebar.
2. The absence of any significant softening in the concrete contacting the top rebar and the maintenance of a tight bond between the concrete and the rebar imply that, for the cathodic protection variables represented in this investigation, rebar bond strength has not been adversely affected.
3. No evidence was found to indicate that the cathodic protection treatments had initiated or aggravated alkali-silica reaction activity in concrete adjacent to the cathode. This, despite the fact that many of the concretes contained rock and mineral types that have, in some instances, shown ASR tendencies. Only one of the concretes examined here showed evidence of alkali-silica reactions. In this case, it was established that the activity was not initiated by the cathodic protection treatment.

Overall Concrete Performance

1. Cathodic protection is being applied to a wide variety of portland cement concretes. In the 15 or so different concretes examined here, cement contents ranged from 560 lb/yd^3 (334 kg/m^3) to over 800 lb/yd^3 (477 kg/m^3) with a wide range of coarse aggregate compositions and maximum particle sizes. A common feature of the concretes examined here is their overall good quality from the point of view of quality of the concreting materials, water-cement ratio, adequacy of the entrained air system (where applicable), and consolidation.

2. The thickness of concrete lying between the cathode and anode in the cores examined here ranges from 1/4 in. (6.4 mm) to almost 5 in. (12.7 cm). In general, this layer of concrete was competent, showing no significant cracking or softening. In those instances where this concrete layer showed significant cracking, it was not possible to say with certainty what role (if any) the cathodic protection treatment played in creating the distress.

ACKNOWLEDGEMENT

A portion of this research was funded by the Strategic Highway Research Program (National Research Council). The publication of this article does not necessarily indicate approval or endorsement by the National Academy of Sciences, the United States Government, or the American Association of State Highway Transportation Officials (or its member states), of the findings, opinions, conclusions, or recommendations either inferred or specifically expressed herein.

REFERENCES

[1] Clear, K.C., "Chloride at the Threshold - Comments and Data." American Concrete Institute Convention Debate, Anaheim, California, Published by Kenneth C. Clear, Inc., Sterling, Virginia, pp. 1-11, March 1983.

[2] Clear, K.C. and Hay, R.E., "Time to Corrosion of Reinforcing Steel in Concrete Slabs, Volume 1: Effect of Mix Design and Construction Parameters." Federal Highway Administration Report FHWA-RD-73-32, April 1973.

[3] Stratfull, R.F., "Experimental Cathodic Protection of a Bridge Deck." Transportation Research Record No. 500, (1974): pp. 1-15. Federal Highway Administration Report No. FHWA-RD-74-31, January 1974.

[4] United States Department of Transportation, Federal Highway Administration, 1991 Report to Congress "1991 Status of the Nation's Highways and Bridges: Conditions, Performance and Capital Investment Requirements." July 2, 1992.

[5] Han, M.K., Snyder, M.J., Simon, P.D., Davis, G.O., and Hindin, B. "Cathodic Protection of Concrete Bridge Components." Quarter Report SHRP-87-C-102B, April 1989.

[6] Hover, K.C., "Cathodic Protection For Reinforced Concrete Structures" in Rehabilitation, Renovation, and Preservation of Concrete and Masonry Structures, American Concrete Institute Special Publication No. 85 (SP-85), pp. 175-208, 1985.

[7] Natesaiyer, K. & Hover, K.C., "Investigation of Electrical Effects on Alkali-Aggregate Reaction in Concrete-Alkali-Aggregate Reactions", Noyes Publications, pp. 466-471, 1986.

[8] Ali, M.G., et. al., "Migration of Ions in Concrete Due To Cathodic Protection Current", Cement & Concrete Research, Volume 22, pp. 79-94, 1992.

[9] Natesaiyer, K. and Hover, K.C., "In-situ Identification of ASR Products in Concrete", Cement and Concrete Research, Volume 22, pp. 79-94, 1992.